JAQUELINE TYRWHITT:
A TRANSNATIONAL LIFE IN URBAN PLANNING AND DESIGN

To Sy

Jaqueline Tyrwhitt:
A Transnational Life in Urban Planning and Design

ELLEN SHOSHKES
Portland State University, USA

LONDON AND NEW YORK

First published 2013 by Ashgate Publishing

2 Park Square, Milton Park, Abingdon, Oxon OX14 4RN
711 Third Avenue, New York, NY 10017, USA

Routledge is an imprint of the Taylor & Francis Group, an informa business

First issued in paperback 2016

British Library Cataloguing in Publication Data
Shoshkes, Ellen.
Jaqueline Tyrwhitt : a transnational life in urban planning and design. – (Design and the built environment series)
  1. Tyrwhitt, Jaqueline. 2. Women city planners–Great
  Britain–Biography. 3. City planners–Great Britain–
  Biography. 4. City planning–Study and teaching–
  History–20th century. 5. City planning–History–20th
  century.
  I. Title II. Series
  711.4'092-dc23

The Library of Congress has cataloged the printed edition as follows:
Shoshkes, Ellen.
Jaqueline Tyrwhitt : a transnational life in urban planning and design / by Ellen Shoshkes.
    pages cm. – (Design and the built environment)
  Includes bibliographical references and index.
  ISBN 978-1-4094-1778-1 (hardback) – ISBN 978-1-4094-1779-8 (ebook) – ISBN 978-1-4094-7374-9 (epub) 1. Tyrwhitt, Jacqueline. 2. City planners–Great Britain–Biography. 3. Women city planners–Great Britain–Biography. 4. City planning–Great Britain–History–20th century. 5. City planning–History–20th century. I. Title.
  HT169.G7T977 2013
  711'.4092-dc23
  [B]
                                                                    2012039503

ISBN 13: 978-1-4094-1778-1 (hbk)
ISBN 13: 978-1-138-24954-7 (pbk)

Contents

List of Figures

Preface and Acknowledgements

I first came across Jaqueline Tyrwhitt when, as a mid-career graduate student doing research in Japan, I was at the home of Catharine and Koichi Nagashima, in Zushi, a seaside suburb of Tokyo. "Have you seen this?" Koichi asked, handing me a copy of the issue of the journal *EKISTICS* dedicated to Tyrwhitt in memoriam; his wife Catharine, Jaqueline's niece, had edited it. I had never heard of either Jaqueline Tyrwhitt or *EKISTICS* before and this discovery was the "aha" moment that brought my dissertation research into focus. Tyrwhitt played a minor, but crucial role in my thesis, which aimed to shed light on the creative interaction of Eastern and Western cultures in the current global transformation, as it played out in the realm of community development in progressive places like Zushi. I was astonished to learn about Tyrwhitt's significant role as an influential transnational actor, generating, disseminating, and cross-fertilizing ideas in this field. The more I learned about her life the more it amazed me, and I determined to try and tell at least part of her story.

The issue of *EKISTICS* dedicated to Tyrwhitt served as my window onto her work and the platform for my research. For that I am deeply grateful to Koichi and especially, to Catharine, who has encouraged and assisted me with this project for many years. Notably Catharine introduced me to her brother, Daniel Huws, to whom Tyrwhitt left her papers. Daniel made available to me a collection of Tyrwhitt's personal papers that were in his custody. This collection, including her diaries, travel journals, and correspondence with Sigfried Giedion, provided invaluable insight into Tyrwhitt's life. I was fortunate to enjoy the warm welcome of Daniel and his wife Helga during my visits to Wales to work with these papers, and to benefit from their personal recollections. I am very thankful for Daniel's continued enthusiastic support and the detailed comments that both he and Catharine made on drafts of this manuscript.

I was able to visit the archives in which Tyrwhitt's professional papers are held thanks to the generous support of the Beverly Willis Foundation; thank you Beverly and Wanda Burbriski. For archival assistance in Britain: thanks to curator Justine Sambrook and her colleagues at the Royal Institute of British Architects Archives at the Victoria and Albert Museum, London, where Tyrwhitt's are held (and her personal papers will soon be deposited); to archivist Grant Buttars at the Patrick Geddes Centre for Planning Studies Collection at Edinburgh University Library; and to Anne Cameron at the Patrick Geddes Collection at Strathclyde University Archives, Glasgow. At Harvard University, thanks to Mary Daniels and Ines Zalduendo for helping me with the Sert and CIAM archives of the Frances Loeb Library, Graduate School of Design. And I thank librarians at the Ford Foundation

Archives in New York and at the Bancroft Library at University of California, Berkeley, where Catherine Bauer Wurster's papers are held. I am very grateful to Alexandros Kyrtsis and the Constantinos and Emma Doxiadis Foundation for enabling me to work at Doxiadis Archives in Athens. Future researchers should know that Tyrwhitt's archives also exist in the CIAM collection at ETH, Zurich.

I have been fortunate to have contacts with people who worked with Tyrwhitt in the World Society for Ekistics (WSE), the Delos Symposia, and Harvard, including Panagis Psomopoulos, the late Suzanne Keller, the late Charles Haar, the late Barry Rae, Irini Sarlis, Marjorie Meyerson, Lawrence Mann, Yvette Mann, Apostolos Doxiadis, Cornelia Oberlander, Marcia Marker Feld, Fumihiko Maki and Jan Wampler. I owe a special debt of thanks to Ray Bromley, who was an early supporter of this project and who drew me into WSE. In addition, there are many colleagues whom I would like to thank for their comments and encouragement over the years that this book has germinated, emerging as a thread in my dissertation and then evolving through several iterations of papers presented at academic conferences and submitted to journals: Susan Fainstein, Salah El-Shakhs, David Listokin, Gilbert Rozman, Christopher Silver, Genie Birch, Pierre Yves-Saunier, Michael Hebbert, Philip Ethington, Robert Freestone, Volker Welter, Daphne Spain, Helen Meller, Carl Abbott, June Komisar and Joe Nasr.

My husband, Sy Adler, has been my partner in every aspect of this project; I am profoundly grateful for his help in the archives, his intellectual collaboration and his acute editing.

List of Acronyms

AA	Architectural Association
APRR	Association for Planning and Regional Reconstruction
ASPO	American Society of Planning Officials
BF	British Fascists
DA	Doxiadis Associates
ETH	Swiss Institute of Technology
GCTPA	Garden City Town Planning Association
GSD	Graduate School of Design (Harvard)
IFHTP	International Federation of Housing and Town Planning
ILA	Institute of Landscape Architects
MARS	Modern Architecture Research Society
MIT	Massachusetts Institute of Technology
NEF	New Educational Fellowship
PACH	Public Administration Clearing House
PEP	Political and Economic Planning
RIBA	Royal Institute of British Architects
SPRND	School of Planning and Research for National Development
SPRRD	School of Planning and Research for Regional Development
TAO	Technical Assistance Operations (United Nations)
TCPA	Town and Country Planning Association
TPI	Town Planning Institute
UIA	International Union of Architects
UN	United Nations
UN-ECOSOC	United Nations Economic and Social Council
UNECE	United Nations Economic Commission for Europe

UNESCO	United Nations Educational, Scientific and Cultural Organization
UNTAA	United Nations Technical Assistance Administration
WFGA	Women's Farm and Garden Association
WLA	Women's Land Army

Part I
1905–1940

Jaqueline Tyrwhitt 1905–1983

Chapter 1
Foundations

Introduction

This book is study of the professional life of Mary Jaqueline Tyrwhitt (1905–1983), a British town planner, editor, and educator who was at the center of the group of people who shaped the post-war Modern Movement. My central contention is that in the course of her work planning for the physical reconstruction of post-war Britain, Tyrwhitt forged a highly influential synthesis of the bioregionalism of the pioneering Scottish planner Patrick Geddes and the tenets of European modernism—the architectural language of housing and social reform—as adapted by the Mars group, the British chapter of *Congrès Internationaux d'Architecture Moderne* (CIAM). The book traces Tyrwhitt's subsequent contribution to the development of this set of ideas—a dynamic merger of Eastern and Western images of the ideal decentralized community, based on cooperation—in diverse geographical, cultural and institutional settings and through personal relationships. This narrative also sheds light on Tyrwhitt's role in the revival of transnational networks of scholars and practitioners concerned with a humanistic, ecological approach to urban planning and design following World War Two, notably those connecting East and West.

Tyrwhitt developed her ideas and her networks in the course of creating new programs to train planners in England, North America and Asia; shaping post-war CIAM discourse on urbanism and assisting CIAM president Jose Luis Sert establish a new professional field of urban design, based on this discourse, at Harvard University (1956–1969); consulting to the United Nations (UN); collaborating with Sigfried Giedion on all of his major publications in English from 1947 on; and helping Constantinos Doxiadis promote a holistic understanding of human settlements, which he termed Ekistics, as a founding editor of the journal *Ekistics* and in the ten Delos Symposia Doxiadis hosted (1963–1972). Those symposia charted the path toward the United Nations' first Conference on the Problems of the Human Environment in 1972, which led to the first Conference on Human Settlements in 1976—Habitat—and forged a worldwide consensus on the concept of sustainable development, and, by 1986, healthy cities.

Tyrwhitt's contributions are insufficiently appreciated largely because she willingly worked as the "woman behind the man." In addition, her work involved translating and synthesizing and mediating ideas that transcended national and disciplinary boundaries, making it a challenge for scholars to see the connections she helped to establish. The book highlights how Tyrwhitt exerted her influence, often anonymously, through collective leadership, or as an intermediary or catalyst. To illuminate such a subtle pattern of influence involves retelling some familiar stories, but from Tyrwhitt's perspective, as she recorded it in letters and journals, as well as her published work. She was not a feminist but this book contributes to feminist literature by exploring the ways in which she worked within the limits of her behind the scenes role, which is often assumed by women in fields dominated by men.

Family Background

Mary Jaqueline Tyrwhitt—known as Jaqueline, or simply Jacky—was a child of the Edwardian period, and keenly aware of her status as a member of one of the oldest families in England. The Tyrwhitt—pronounced Tirrit, "as in spirit," Jaqueline would remind people—family tree goes back to Sir Hercules Tyrwhitt, said to have been knighted by William the Conqueror in 1067. As descendants of the original English gentry, "who remained as of old without seeking wealth in trade," the younger sons of the family, who did not stand to inherit money, entered the church or the law (Tyrwhitt 1862: iii). Jaqueline's father, Thomas (1874–1956), was the son of Reverend Henry Mervyn Tyrwhitt (1844–1937) and Jaqueline Frances Otter (1849–1936), whose father, Rev. W. B. Otter, was the Archdeacon of Lewes. Thomas became an architect. Following several years of apprenticeship, study at the Architectural Association, and a year at Oxford in 1898, he opened his own practice in 1901. However, he promptly abandoned that enterprise to seek his fortune in Hong Kong, then in the midst of an economic boom, where he worked for a year. Thomas returned to Britain to marry Dorothy Nina Marsden (ca. 1883–1938), the daughter of a lawyer, in February 1904. Thomas and Dorothy immediately moved to South Africa. Jaqueline was born in Pretoria, South Africa, on May 24, 1905.

In Pretoria Thomas worked as an architect in the Public Works Department where he contributed to designing schools, post offices, and other government buildings—part of the reconstruction effort after the South African (Boer) War of 1899–1902. His work, designing model plans for school buildings to standardize arrangements and staffing, exemplified Progressive era reforms aimed at modernizing civic infrastructure. At the conclusion of his two-year contract Thomas moved his growing family, which now included a new baby girl, Edrica (1907–1999), back to London.

Thomas and Dorothy settled in Hampstead, north of London. Jaqueline's brothers, Robert and Cuthbert, were born there in 1909 and 1912 respectively, as was another sister, Alicia, in 1916. Thomas opened an office in London, but his fledging practice was interrupted in 1914 by the outbreak of World War I. He promptly joined the Royal Naval Voluntary Reserves and served on the home front in the anti-aircraft corps from 1915 to 1918. Young Jaqueline's sense of family pride and patriotism was most certainly fueled both by her father's military service as well as the decorated heroism of her most famous living relative, Commodore Sir Reginald Yorke Tyrwhitt (1870–1951), who commanded Britain's Home Fleet flotilla during

Figure 1.1   Thomas Tyrwhitt with Jaqueline as a child, ca. 1906
Source: © Constantinos and Emma Doxiadis Foundation

the war. The connection between Jaqueline's immediate family and Sir Reginald was strengthened when her aunt, Ursula Tyrwhitt (1872–1966), married Reginald's older half-brother, Walter Spencer Tyrwhitt (1859–1932) in October 1913. Ursula resisted marriage until she was 41 in order to pursue a career as an independent painter; she was one of the early women students at the Slade School of Fine Art in London (Thomas 1996, Wilson 2003). Jaqueline [E.1] described her mother as "an enthusiastic follower of the teaching of Octavia Hill," the Victorian anti-suffrage reformer who "embraced female activism in the realms of church, charity, and education" as acceptable spheres for women's work (Ross 2007: 19).

Thomas Tyrwhitt did not immediately return to his private practice after the war, but instead contributed to the reconstruction effort, accepting a position with the Ministry of Agriculture and Fisheries to oversee the design of 32 cottages as part of an experimental farm settlement. One of the objectives of this project, conducted in conjunction with the Department of Science and Industrial Research, was "to establish an entirely new standard for working-class housing"—in response to post-war demands for better living conditions. Another objective of the project was to examine "the practicality of reviving the use of local materials and traditional [rural] building methods, as well as the usefulness of certain new construction methods," such as cob, concrete blocks, chalk and cement blocks, timber and *pise de terre*. [T.1] Country councils could build the houses with government funds, as authorized by the Housing and Town Planning Act of 1919. "In view of the present housing difficulties," reported the journal *Nature* (Experimental 1920: 792), there was "considerable interest ... centered in the results of the[se] experiments in cottage building."

Thomas resigned from his position at the Ministry of Agriculture and Fisheries in June 1920, however, when he garnered his first commission by winning the design competition for the Indian Memorial Gateway at the Royal Pavilion in Brighton. Jaqueline, then 16, proudly attended the unveiling of her father's creation by His Highness the Maharaja of Patiala (India's 1921: 7). Thomas hoped this high profile project would lead to more work. Like his sister, he cherished his independence and preferred to earn his living through creative work, rather than in a salaried job.

This was the cultural context of Jaqueline Tyrwhitt's childhood: somewhat sheltered, yet cosmopolitan; somewhat privileged, yet unstable economically while her father struggled to establish himself professionally; interrupted by a horrific war, yet brightened by a bit of the afterglow of Edwardian society, which was prolonged by the post-war boom. The family's finances suffered during the economic slump that followed the boom. Thomas, a shy man, was not good at self-promotion, and his practice did not flourish. Moreover, he was frugal. Jaqueline helped her mother care for the younger children and with household chores; she felt keenly aware of the family's genteel poverty.

St. Paul's School Days

Jaqueline was fortunate to win a scholarship to attend the prestigious St. Paul's Girls School. The school itself was relatively young when she enrolled as a day student in 1918, having been established in 1905 by trustees of the venerable St. Paul's Boys Schools, itself founded in 1509. She showed an early affinity for gardening and formed a close bond with the school's noted gardening mistress, Chrystobel Procter (1894–1982). In her fourth year at St. Paul's, Procter entrusted Jaqueline with responsibility for planning the school's market garden, and supervising the work of younger students. However, she received poor grades in her other classes that year. St. Paul's encouraged students to continue on to universities, but Jaqueline confided to her diary [4/22/22] not knowing "what to do about matric to start & then Oxford." She learned in late July that she failed her matriculation exams.

That summer Thomas sat down with Jaqueline to discuss her future. She would need to earn her living and he decided that her career would be in gardening. He arranged for her to spend a fifth year at St. Paul's under the tutelage of Procter, to prepare for the general horticultural examination of the Royal Horticultural Society. Her future course of action now clarified, Jaqueline buckled down and passed her matriculation exam in October 1922. The next term she gave up piano lessons—she was a talented musician—in order to concentrate on botany, chemistry, and horticulture. She probably used as texts Arthur Tansley's *Elements of Plant Biology* (1922), which contained an early formulation of the ecosystem concept, and Tansley's *Practical Plant Ecology* (1923), the book that introduced plant ecology into schools at that time (Anker 2001); both were enduring influences. Jaqueline passed the Royal Horticultural Society's general exam with high marks in March 1924.

The next logical step would have been for Jaqueline to continue her training at Swanley Horticultural College, then the top private gardening school in England. But her family had to pay the tuition and fees for her brother, Robert, the eldest son, to enter a private Naval College in order to pursue the life of military service charted out for him. Swanley was now out of the question for Jaqueline because it was too expensive.

Jaqueline's father came up with an alternative that set her on a different trajectory. "He and I first pulled my career to bits and pieced it together again & then went off to interview the Principal of the Architectural Association ... . I am to become a student for a term in order to learn about plan drawing & something about the history of architecture & the various styles. It will be great fun but I fear rather costly." [9/18/24] Jaqueline worried that her mother was not pleased; attending daily classes meant she would not be available to help around the house and with her youngest sister, who needed constant care. Yet Dorothy apparently agreed that her daughter's education was worth this sacrifice. To earn some money, Jaqueline planned to do weekend work on gardens belonging to various relatives and family friends.

At the Architectural Association

Jaqueline was among eight women out of 30 students who joined the first-year class at the Architectural Association (AA) in September 1924. The AA first opened its doors to women in 1917; this decision was both in response to the loss of male students during the war, as well as to the success of campaigns for women's suffrage and employment. Training at AA, firmly following traditional Beaux Arts principles and methods, focused on drawing until the late 1920s. At first, Jaqueline found the discipline of geometric drawing to be a bit maddening, but she immediately enjoyed sketching statuary in the nearby British Museum, which became one of her haunts. Entering students stayed together as a class; each "year" worked together in a studio, a large room where students set up their drawing boards around shared tables. Jaqueline enjoyed studio life tremendously. The time Jaqueline spent at AA also brought her closer to her normally reticent father. He accompanied her to events at the school and she developed a deeper appreciation for his work, which they discussed. She was proud that he was a Fellow of the Royal Institute of British Architects and known to the other students.

Service: Girl Guides

Even as Jaqueline was exposed to new people and ideas at AA, she maintained allegiance to her old school friends ("Paulinhas") through her membership in the Girl Guides. She had joined the Guides during her last year at St. Paul's and served as a patrol Guider (leader) at Dame Colet House, the settlement house St. Paul's ran in Stepney, in London's East End. Jaqueline took pride in her family's service to the Empire—which reached its zenith in the 1920s—and found, in the Girl Guides, a way to serve in her own way. The Girl Guides had overtly militaristic overtones when Agnes Baden Powell formed the organization in 1910 as a counterpart to the Boy Scouts. By the 1920s, however, the emphasis in the Guides' movement had shifted from "Imperial defense … toward training citizen-mothers in a new international environment … to help rebuild the war-torn world" (Procter 2009: 35–6).

Dorothy Tyrwhitt probably encouraged Jaqueline to join the Guides. But it appears as if her Aunt Vere—who knew about "the lower classes – slums & hospitals & so on"—inspired her to continue as a Guider after graduating from St. Paul's. Vere was the wife of her mother's brother Edward Marsden. Both of her mother's brothers, Edward and Hugh taught at Eton College, the famous public school. "I wish I could be here at Eton for a month or so – it would do me all the good in the world," Jaqueline wrote after a weekend there in December 1924. "Aunt Vere is really awfully well up unemployment etc & she has the power of looking at things from their point of view really & truly," the nineteen-year-old Jaqueline wrote in her diary [12/7/24]. She did not, however, share her aunt's perspective:

> I am most awfully interested in my Stepney girls as human beings of a fascinating type – but the squalor, sordidness – the hand to mouth grabbing down there can't appeal to me. If I lived there I'd go mad. I just couldn't see the beauty through the dirt – humanity in the mass revolts me & I long for nature where I belong.

She loathed having to wear her "full Guide rig" all day; at the AA, she ate "lunch hidden away in the cloakroom." Jaqueline's ambivalence about identifying herself as a Guide at the AA belied her true feelings. "Stepney! I wish I could enjoy slumming. I don't. I do it [because] I once let myself in for it & now I can't get out *with honour*." [1/8, 22/25]

Jaqueline's year-end reflection for 1924 is worth quoting at length, as it articulates her mixed sense of accomplishment and anticipation, confidence and conflicted ambition, and personal insecurities as she is poised to leave her teenage years behind:

> I have seen a great many different people & their various modes of living though I have made no new great friends. … [M]y true desire & vocation is not to follow any profession (tho' I do love that which I have chosen) but to marry & beget children. Added to that is an ever growing love of freedom & an equal ever growing horror of marriage restrictions & household duties. I long to travel – to travel widely. Oh I wish, I wish I was a man! I want to go alone – to poke about in all sorts of odd quarters of the Globe. I want badly to earn my living – to be worth my salt – to try to live down my Secondrate Nonentityship – but yet I do not want to grind & the monotony of it! I have no irrepressible urge to design gardens though I love to do so. However I think a regular occupation is the only thing for me … [1924, memorandum]

Gardening with Miss Willmott at Warley Place

Jaqueline also maintained contact with her Paulinha friends at weekly rehearsals of the Bach Choir. That is also where she met her mother's friend, Ellen Willmott (1858–1954)—whom none other than Gertrude Jekyll called "the greatest of all living women gardeners" (Brown, 1999). Willmott took an interest in Jaqueline and invited her to attend lectures at the Garden Club. Jaqueline learned a lot from those talks, and also enjoyed being known as Miss Willmott's protégé. In late February, with the end of her term at AA a month away, Jaqueline looked to Willmott for career advice. "She is supposed to be planning it out for me!" [2/24/25]

Willmott proposed that Jaqueline work with her at the Amateur Gardeners Association, which was affiliated with her estate Warley Place. The prospect of a summer in the country, working in Miss Willmott's famous garden—considered "not only one of the most beautiful, but also one of the most interesting of English gardens" (Misc. 1934: 398)—was exciting and daunting. "Lord I shall have to be meek & humble [because] she does know so much & I am so bad at working in with authorities!" Jaqueline acknowledged. [3/13/25]

In 1925, Willmott was enjoying a respite from many years of struggle to hold onto her estate—a brick mansion and 33 acres that her father purchased in 1875— which she had inherited along with a fortune, in 1902. She exhausted her wealth by expanding her holdings and developing her celebrated gardens, largely before 1914. She applied a scientific approach to her gardening and was renowned for her experiments in cultivation and hybridization. Willmott's knowledge gained her entrée as a member of the male-dominated Royal Horticultural Society (RHS). But her spending nearly bankrupted her. By 1921 she had secured Warley by selling off two gardens she owned abroad. When Jaqueline arrived, Willmott was starting over—resuming her experiments, her committee work at RHS, and judging rose shows—while reorganizing her finances (Le Lièvre 1980).

Soon after Jaqueline settled into her work for the Amateur Gardeners Association, which probably handled Warley's seed operation, Willmott offered to take her on in the garden too. This was an honor: she was one of the few women Willmott trained. Yet Willmott sorely needed Jaqueline's help; out of a staff that once included over a hundred gardeners, only Jacob Maurer, there since 1894, his son, and a few others remained (Le Lièvre 1980).

Jaqueline was thrilled to work alongside Willmott. "I am so happy here that I'd love to stay here a whole year round," Jaqueline wrote soon after her arrival at Warley.

> I believe that ... Warley Place is a new Eden & the house itself is so full of lovely, rare, wonderful & beautiful things that I rejoice whenever I enter it & fill my eyes with beauty. Miss W herself is of course the gem of the collection. Full of kindliness & fun – deliciously wicked & fantastically generous – full of whimsical economies and reckless expenditure .... Tho' I am perhaps impertinent I admire her whole heartedly. [4/30/25]

Jaqueline also enjoyed working with the knowledgeable yet often dour Jacob Maurer. In fact, Jaqueline's experience at Warley brings into focus for the first time her ability to appreciate and work with a wide variety of people.

When Jacob's son left Warley, Jaqueline volunteered to take on his work in the Alpine garden. This meant doing her job for the Amateur Gardeners' Association at night. By mid-summer Warley no longer seemed so Edenic.

Seeking a Political and Spiritual Ideal

Despite working hard all week at Warley, Jaqueline continued to serve in the Girl Guides. She now maintained this commitment largely because, in March 1925, she had joined the British Fascists (BF). "I don't want to cut off from Stepney espec. if I am to become a British Fascist." [3/19/25] Rotha Linton-Orman (1895–1935), who founded the BF in 1923, was among the first group of Girl Guides and drew on that experience along with what she knew of Mussolini's experiment and her own military background. She formed the BF as a paramilitary organization aimed at assisting the government in the event of a civil emergency, such as a general strike, as was then being threatened by organized labor. BF, which targeted women in its propaganda, advocated violence only if necessary to maintain social stability; its members were middle- and upper-middle class—respectable, not rowdies, anarchists or revolutionaries. BF members were encouraged to vote for the Conservative Party in elections; otherwise, the group's policies were vague and super-patriotic: supporting a strong empire and monarchy, and opposing immigration and socialism (Gottlieb 2000, Farr 1987, Thurlow 1987, Linehan 2000).

Jaqueline joined BF at the peak of its membership, having been recruited by a friend from AA. While her motives are unclear, Jaqueline's association with BF merits examination in order to define a baseline position from which she undertakes, over the course of the next 15 years, an intellectual and personal odyssey—involving the transformation of her political ideas from ultra-conservative to social democratic.

Jaqueline's diary illuminates how she fitted membership with the BF, along with the Guides, into her schedule while living at Warley. For example, on May 16 1925:

> I got up at 7.0 … was up at Warley by 8.0 – picked a quantity of bluebells etc & caught the 8.49 from Brentwood! Then down to Stepney – arrived there by 9.30 – and then … home to change for Chelsea [garden show] & to Bute [St. Paul's garden] to pick up C[hrystabel] P[rocter]. The show was splendid & I enjoyed roving around with C.P. ever so much …. After Chelsea – i.e. 6.30 or so I returned home & found a British Fascist notice – bounced out again & interviewed a Company Commander & the District Commissioner & his wife & learnt lots more. Bought my badge among things. Then dinner & then Garden Club …. Then back to Warley … and its now late.

A subsequent entry captures Jaqueline's childlike sense of adventure in attending a BF rally with her teenage brother Cuthbert, while also hinting at a religious dimension of her attraction to the Fascisti—among whom a fervent Christianity was central to loyalty to king and Empire (Jackson 2010).

> I went to Mass at Westminster Cathedral & somehow, altho' I loved the music I did not find God. Then lunch at Granny's …. Then a Fascist Meeting in Hyde Park. I went with C[uthbert] & enjoyed it tho' I didn't dare go in & form up with my Company [because] I promised Mother to get back. We had to leave just as it [became] fun & I did not meet … [friends from the AA] or anyone I knew. [5/24/25]

The Tyrwhitt household was devout. Dorothy was High Anglican, and many men on both sides of her family had made careers in the Church of England. But as Jaqueline approached the age of 20, she was looking for her own spiritual home. In her search she attended a variety of Christian churches. She blamed herself for being too demanding: "I need such a lot from my religion which is why I can't find one." [4/12/25] While living in Warley, she began a practice of worshipping at the local Catholic Church during the week, and at the Church of England on Sunday. This combination provided both the warmth and the clarity she was seeking:

> I enter the C of E – stern unheard dictums compel to my knees & I recite the Lords prayer perfectly word for word – no more …. I enter the little Catholic church. I murmur the Lords Prayer to a tactfully sympathetic guardian and

light my little candle with a childlike glow of faith. But on Sunday the C of E is become a place where one can commune with God in the simplest and most dignified of service – full of beautiful English and lofty ideals – And now the R.C. church is become a Pagan Temple where sensual carnal worship is practiced – full of ancient rituals & sexualism. [5/11/1925]

Jaqueline's turn toward Catholicism was part of new traditionalism in religion in Britain, which rose in response to new uncertainties of science, and doubts about Enlightenment ideals in the aftermath of World War I and post-war economic malaise (Mowat 1967). For ultra-conservatives such as the BF, who sought a spiritual renewal of the modern world, an ideal Christian society, the religious and the political expressions of these ideas were conflated (Jackson 2010). However, Jaqueline kept her interest in Catholicism—and spiritualism—to herself. Both were "subjects that are regarded with grave suspicion & fear – even horror – by almost all my relatives and friends." [7/26/25] At this time, Jaqueline felt increasingly at odds with her parents. "I do love honour & respect my parents – but quite impersonally … . I have fought with Mother many a time – I am always quelled eventually. I have argued with Daddy – I am always proved wrong, ignorant & uppish. They unite in telling me that I am the only one of the family that behaves like this." At the same time she was anxious about her family's "impecunious state." [9/6/25, 10/25/25]

In November 1925, while staff at Warley were busy preparing for a planned visit from Queen Mary, a fascist rally in Hyde Park drew an estimated 10,000 people (British 1925). As a BF member she would have been informed of intensifying preparations in the event of a general strike as the May 1 deadline approached. In April, the Home Secretary declared that it would not cooperate with BF unless the organization explicitly committed itself to parliamentary democracy. This forced a major split in the group. Lintorn-Orman and the women on the general council who refused to give up the fascist name and paramilitary structure now took over the organization's leadership (Gottlieb 2000).

In many ways the general strike, which began on May 4 and lasted nine days, marked the end of this phase in Jaqueline's personal and intellectual development. There is no record that she engaged in any action during the strike, although her father joined many others in becoming a special constable. Politically, Jaqueline began to drift away from the fascist movement—which at any rate had lost its steam at least in this incarnation—and towards the Conservative Party. She also began to drift away from gardening.

Chapter 2
New Landscape, New Society

Coming of Age

Despite her fondness for Ellen Willmott, the hard work, meager pay, and lonely life as an apprentice gardener led Jaqueline to leave Warley Place after less than a year. Having disappointed her parents, Jaqueline moved in with her more sympathetic maternal grandmother, who lived in nearby London. By mid-1926, Jaqueline had found a new job working for a small firm of "garden architects" conveniently located in the neighborhood. At the same time she enrolled in an evening course at the London School of Economics (LSE). Thus began a new phase of her intellectual development, involving a widening of her focus from garden design, botany, and horticulture to include social economics—or in other words, human ecology.

As Jaqueline turned 21 she also abandoned her youthful flirtation with fascism and became active in the right wing of the Conservative Party. Her attraction to party politics was probably spurred as much by fragmentation of the BF after the failure of the general strike as by passage in 1928 of the Representation of the People Act, which extended the franchise—already granted to propertied women over 30 in 1918—to women over 21 who met property qualifications. All political parties in Britain mobilized to compete for the expanded electorate, especially young women, "the flapper vote" (Mowat 1967). The Conservative Party led the way in organizing to attract women (who outnumbered men voters by two million.) By April 1929, Conservative Party organizers had established hundreds of branches of its coed youth organization, the Junior Imperial League (JIL) (Heathorn and Greenspoon 2006: 95). Jaqueline joined her local branch in Hammersmith.

Jaqueline established significant relationships through the JIL. Her "best friend" was Stephen Bull (1904–1942), a young solicitor whom she worked for informally as a "private secretary" in his capacity as Vice Chairman of the all-male Hammersmith South Conservative Association. Stephen was the eldest son of Sir William Bull, First Baronet (1863–1931), a leading right-wing politician who had represented Hammersmith first on the London County Council (LCC) and then in Parliament for 37 years. Jaqueline found in Sir William a father figure at a time when she felt estranged from her parents. She valued his "wisdom, guidance, sympathy – real sympathy – & friendship." [1/27/31]

Sir William Bull advised Jaqueline, early in 1929 amidst rising unemployment, to leave her job, where as the sole surveyor/designer/draftsman she felt overworked and underpaid. He supported her in this move by hiring her to run the office of the Imperial Society of Knights Bachelor, a charitable organization he directed. He also supplied crucial emotional support when Jaqueline experienced a personal crisis in late December 1929, triggered by the death of her grandmother. Sir William helped Jaqueline come to terms with her grief, and probably convinced her to reconcile and move back in with her parents.

Jaqueline applied herself energetically to her job for Sir William, which included the publication of an annual directory of the Knights. The offices of the Society were in the same building as Sir William's law practice, which facilitated interactions with both Stephen and his younger brother George (1906–1968), who were involved in the work of the Society as well as the law practice. Her office served as the hub of her increasingly interconnected social, professional, and political networks. The job allowed her substantial autonomy and opportunities to take creative initiative. [T.2] This was a time and place when Jaqueline honed her understanding of political ideas and developed a valuable new set of administrative and political skills, not least of which was the art of cultivating influential friends.

"Progressive" Toryism

Jaqueline had high regard for Sir William Bull's "breadth and strength of vision." A so-called Tory "diehard," Bull was a staunch imperialist. But he also advocated on behalf of several progressive initiatives. He was one of the first (in 1901) to propose a regional planning solution to the problems of London's sprawling growth, namely a green belt to contain development and protect the countryside (Thomas 1963: 14). Bull also championed a role for government in developing transportation and communications infrastructure, such as a tunnel under the English Channel, and chartering the British Broadcasting Corporation (BBC) in 1926 (Mowat 1967: 242–3).

Bull's "conservative socialism" provided a precedent for the ideas of a group of Young Tory progressives anxious to provide their party with a constructive alternative to socialism. "[T]hey sought a 'middle land' made of the both of two extremes … socialism and laissez-faire." (Ritschel 1997: 39). The Young Tories proposed mergers, trade associations, and other forms of corporatist industrial organization to protect British domestic industry from competition, and they called for using the power of the state to aid the workingman as well as industry (Mowat 1967: 347). Jaqueline received a personal tutorial on these ideas from her friend John Hoffham "Jack" Blaksley, whom she probably met through political circles around the time—spring 1929—that the *National Review*, a leading right-wing journal, published his essay, "The Tory Ideal." They soon forged an enduring friendship although they made an unlikely pair. Blaksley, a retired army captain, was 20 years older than Jaqueline and married.

Blaksley's (1929: 540–44) conception of the Tory ideal resonated with values Jaqueline had embraced as a Girl Guide and British Fascist: "the higher moral values in life, for those traditions of honour and loyalty of duty, responsibility, and obedience, of reverence, restraint, and self-respect, upon which the British character is based." In leaving garden design behind, Jaqueline embraced the cause of the Young Tories along the lines Blaksley defined:

Our way lies in a better individualism … [and] the fulfillment of that right to opportunity which is the basis of all true social reform … .We believe in local autonomy as opposed to centralized bureaucracy, and we see in service to the locality – the village, the township, or the county – the nursery, as it were, of those larger ideals of service to the country and the Empire which national patriotism implies.

Extended Family

Jaqueline befriended Blaksley at a time when she was feeling adrift personally and professionally. Her beloved great aunt Mary passed away. Jaqueline had turned to Mary as well as her recently deceased grandmother as a refuge from her troubles. In grieving for them, she also mourned the loss of their shared delight in "the pleasures of life," which her "stoic" parents did not attach importance to. [1/18/31] The end of a serious, albeit brief, love affair added to her despair. These personal losses upset the core of her world and reverberated the more generally unsettling effects of the worldwide depression.

Jaqueline found a supportive community through her political activism, which happily reinforced her connection to the Bull family. In January 1931, she threw herself into JIL committee work, fielding candidates to stand in the upcoming LCC election. She enjoyed plotting strategy with Stephen Bull, devising her tactics at JIL to achieve his Conservative Association goals. To play such a behind the scenes supporting role to a man suited Jaqueline's ambivalence about her identity as an independent woman. However, her success in those maneuvers bolstered her self-confidence, and she resolved to put herself forward to become the JIL secretary in the coming year.

Death of Sir William Bull

The sudden death of Sir William Bull in January 1931 was a devastating blow to Jaqueline. Her initial response was concern for Stephen, and then she feared for her own fate. "I don't know what may befall me – I don't know where I am – I am lost & alone." Overwhelmed by "bitter grief and self pity" and feeling as if she were being punished for her independence, Jaqueline beseeched God:

Why do you force me … to acknowledge my parents as my omnipotent rulers? … Had I followed Daddy's wish & stuck to garden designing depending for my sustenance on gratuities from Mother I should never have taken on this job – & Sir William's death would not have meant the crash of all my life – and yet God I defy you. While I have breath I will not submit to them. I will be free – free of my parents – but please God be merciful. I do not want to be free and alone – not alone God –grant me some comfort. [1/25/31]

The Bull family immediately reached out to Jaqueline, inviting her to their home the weekend of Sir William's death. She realized she was there "not so much to discuss details" as to provide a familiar, steadying presence, much as Sir William had done for her when her grandmother had died. She was grateful for the opportunity to be of service to this family she loved. Gradually, she took on responsibility for the church seating arrangements, procuring a hatchment of his ancestral arms (an ancient knightly custom she knew Sir William would have wanted), and enlisting ushers for the funeral. A "huge crowd" attended the memorial service including many dignitaries. [1/29/31] In effect, Jaqueline served as a stage manager for this occasion, revealing an innate understanding of both the pageantry and protocol required for its success.

After the funeral, Jaqueline reflected on what her next move should be, aware that she had depended so "whole heartedly" on Sir William. Her first instinct was to "get right away" but "now oh God I have a terror a horror of the unknown." Nevertheless, she resolved: "I will not be dependent on my parents and I will not marry a weakling – and these are my only two obvious ways out." To marry Stephen Bull—whom she loved "tenderly and more dearly than [her] life"—was not an option. She had long known that "as his 'confidential private secretary' [she was] a valuable – an integral part of his life – but as his wife – the idea [was] out of the picture." [1/29–30/31] Apparently Jaqueline was learning that playing a supportive role was also a good strategy for maintaining an intimate and vital connection to a man loved platonically (or unrequitedly). "If only somewhere I could be needed … If I were a necessity – a valued necessity – I would slave to the bone" she vowed. [2/11/31] This capacity – and, perhaps, need – for self-sacrifice was one of Jaqueline's salient characteristics.

National Emergency

In the short term, Jaqueline threw herself into political activism where she found a place for herself in the world outside her parents' home. In her new role as JIL secretary, Jaqueline relished taking part in debates on domestic and imperial affairs, mock elections and parliaments, and guest lectures. Her fears about her own future melded with her broader concerns about the "national emergency," and the future of the Empire (particularly India, a topic high on the Conservative agenda). Jaqueline longed for a leadership change: "Winston [Churchill]'s the only genius – & no-one's going to trust Winston although he is a man of brilliance & a leader." [2/4/31]

At this time Britain's economy was rapidly deteriorating; this in combination with other factors such as fear of communism and the advancement of science, created a climate of opinion that encouraged interest across the political spectrum in the concept of national government planning as a solution to what was becoming widely perceived as the breakdown of market capitalism. Labour Party leader, Sir Oswald Mosely, issued a call for national planning. In response, the "dissident Tory" journal *Week End Review* (*WER*) published its proposal for "A National

Plan for Great Britain." Written by a contemporary of Jaqueline's, *WER* assistant editor Max Nicholson (1904–2003), this plan built on the earlier ideas of the Young Tories in calling for a reorganization of industry based on a corporatist model of self-governing industries. Nicholson, who had studied ornithology and zoology, conceived of the proposal for "planned capitalism" in ecological terms: as "a workable basis for industrial civilization other than fascism or communism" explicitly formulated to unite "all progressive elements in the country behind the demand for planning" (in Ritschel 1997: 147). Leading industrialists, such as auto manufacturer Sir William Morris (later Lord Nuffield), supported this approach (Andrews and Brunner 1955). Morris had founded the National Council of Industry and Commerce (NCIC) in 1930 to lobby for a nonpartisan, technocratic approach to government: applying scientific principles of management along business-like lines (Thomas 1959: 3). Jaqueline was impressed to learn from Blaksley, the Tory ideologue, that he had taken a job as an NCIC organizer.

League of Industry

Blaksley kept Jaqueline abreast of NCIC, which in April 1931 changed its name to the League of Industry. In June, Blaksley arranged for the League to hire her to work as his chief assistant in the organization of the southern and western areas of England. This brought an end to Jaqueline's active involvement in local Conservative Party politics, as the League was purportedly a non-political industrial movement—albeit one firmly in support of Conservative goals of protective tariffs and imperial preference.

Right after Jaqueline began her new job, in July 1931, an international banking crisis precipitated a domestic crisis that brought down the Labour government, and led to the formation of the National Coalition, which was supposedly a "national effort in cooperation" but in reality was Conservative (Mowat 1967: 399–400). Jaqueline's work for the League increasingly focused on industrial peace. "What the country wanted was a form of the old guild trades union associations, which looked after their trade and did not dabble in party politics," League chairman and industrialist Sir Wyndham Portal reported at a mass meeting in Nottingham in January 1932. Another League official asserted that employers and workers increasingly realized that "only an efficient industrialism"—not class war—could bring the social and material benefits they sought (League 1932: 14).

Political and Economic Planning

Meanwhile the *WER* plan inspired the creation of Political and Economic Planning (PEP), which was organized in March 1931. The conveners were two of Nicholson's professors at Oxford: zoologist Julian Huxley and biologist Alexander Morris Carr-Saunders; they invited those who had worked on the *WER* plan and members of a like-minded study group "to carry on an independent non-party discussion and study this whole range of public affairs" (Nicholson 1981: 8). For financial support, the PEP executive committee turned to the Elmhirsts: Dorothy, a wealthy American philanthropist, and Leonard, a Yorkshire-born, Cornell-trained agronomist. At a meeting in April 1931 at their home, Dartington Hall—a medieval estate in Devon, which the Elmhirsts were rehabilitating as an experiment in rural revitalization—they agreed to finance the organization for the first year, and Leonard become an adviser (Lindsay 1981). The leading roles of Huxley and Carr-Saunders—and now Leonard Elmhirst—signaled the expansion of ecological concepts into the realm of economics at this time (Anker 2001).

PEP held its first general meeting in London in June 1931, just as Jaqueline began to work for the League of Industry. Those attracted to PEP were mainly men in their thirties and forties, to the right of center or the left-wing of the Conservative party (Ritschel 1997: 152). Initially membership was limited to around 50, tightly organized in groups, each dealing with a component of the "National Plan" to be produced: agriculture, land use, economics, government, industry, and the technique of planning. Jaqueline had ties to the industry group through its secretary Oliver Roskill. Jaqueline had known Oliver since childhood; she considered his mother as her "ideal of womanhood:" "all fire and intelligence—a strong & forceful personality with firm and reasoned convictions—also an eloquent & attractive speaker but never a 'public woman.'" [2/16/31]. However PEP spent most of its first two years "trying to get down to work" and being overwhelmed and preoccupied with the "search for a definition of planning" (Ritschel 1997: 158–9).

International Travel: Burg-El-Arab Egypt

No doubt it was her work for the League of Industry that led Jaqueline to take advantage of her paid vacation, in February 1933, to go to Egypt to visit Wilfred Bramley-Jennings (1871–1960) and his family in Burg-El-Arab, a town he built for the Bedouins 40 miles west of Alexandria. Jaqueline was probably encouraged to visit Wilfred—known as Wiffy—by his mother Bertha Bramley (1948–33), for whom Jaqueline had designed a garden in 1925, and whose father had been British Consul in Alexandria [1/10/31]. Jaqueline kept a journal which she intended to share with Blaksley who had fought in Egypt during the First World War. [3/1/33] She began it by recounting a meeting she and Blaksley attended, just before her vacation, at Welwyn Garden City. Bramly's plan for Burg-El-Arab owed something to the spirit of the garden city movement as well as to its predecessors, model industrial villages of the late nineteenth century, and the crafts cooperatives promoted by John Ruskin (1819–1900). It also exemplified the cooperative industrial ideal promoted by the League of Industry, PEP, and Dartington Hall.

Jaqueline recounted: Wiffy had received a grant from King Fuad to build a desert town that would function as the headquarters for desert administration officials, a place to meet with local Bedouin tribes. The walled town included a

Courthouse, shops, and a market place. In 1919, Bramly had the idea to build a carpet factory to help Bedouin war widows. "400 Arab women were trained in their native craft …. Wiffy had visions of founding a new school of rug making … & he dreamt of Burg rugs holding the same status as Bokhara rugs." [2/24/33] The scheme worked well until the funding stopped in 1924. Bramly was forced to abandon his plans. He then devoted himself to building his house, Dar-al-Badia, on a ridge overlooking the town.

Knowing of Wilfred's interest in gardening, Jaqueline brought seeds with her to plant in the ancient quarry he used for a "wild garden." She intended to bring home as much seed as she could (following in the famous seed collecting footsteps of Willmott). Jaqueline planned to "diagram … the habit & environment of each plant" and reflected: "It is so tremendously restful & pleasant to be again with growing things—flowers, paints, buildings—all that side of life that is, necessarily, completely hidden 'on the job.'" [2/24–3/10/33]
Jaqueline's travelogue illuminates her probing curiosity and a mode of thinking that could be described as human ecology:

> A large part of my conscious mind in the desert was employed in observation – the light on the hills, the flowers underfoot, the mirage beyond and the animals & people one encountered … . In observing and imbibing their beauty and also in reasoning out their cause – whether the light was caused by direct rays of the afternoon sun or by reflection; what family the flowers belonged to, how they came there and how they continued to live – their natural adaptations, hereditary and self made; the causes of the mirage and the mode of life of the animals and people. [3/10/33]

While in Egypt, Jaqueline did not, as she had planned, give much conscious thought to her "future mode of life." Nevertheless, "unconsciously things have dropped into perspective," she believed. She resolved to maintain her "present mode of life" but "study more of the background of the job (economics, history & geography) and take more active personal part – and a certain amount of speaking &, perhaps, writing." At the same time, she would "continue & expand my social life always seeking (but not in active pursuit of) 'useful' employment." She wanted "to do something worth while – to justify myself." But she would have to contribute "through human organization in some form or other – not, alas, through artistic endeavor." For this, there were "few jobs that could be a better training" than her current one, if she took it seriously." [3/10/33]

Industrial Reorganization and Regional Planning

Shortly after Jaqueline returned from Egypt, in May 1933, the League of Industry invited PEP to speak at its second annual conference later that year. PEP viewed the League of Industry as a natural ally, "ripe for infiltration," and readily accepted

Figure 2.1   From Jaqueline's journal 1934, photos of Burg-El-Arab, Egypt

the invitation (Ritschel 1997: 179–82). League organizers like Jaqueline probably received PEP's second Broadsheet (May 1933) on Town and Country Planning. By then a new approach was emerging within PEP work groups—planning for the location of industry as part of regional development planning. These ideas resonated with those being discussed by the League's membership, and they adopted two planning-related resolutions at the conference: one concerned employment and the other supported a central planning authority for agriculture (Planned 1933: 11). Jaqueline's work for the League of Industry was now closely aligned with the PEP Industries Group, which brought her closer to Oliver Roskill. Through Roskill she became an active member of the Industries Group which was among the most active in PEP's formative period, 1933–1939, when PEP became a leading force in the formation of the "middle way" consensus.

International Travel: Shanghai via Trans-Siberian Railroad

Roskill took Jaqueline out for a night on the town as a send-off for her next big adventure: a journey to Shanghai via the Trans-Siberian railroad in March 1934. Jaqueline had saved her annual vacation time in order to visit her brother, Cuthbert, a junior army officer whose battalion had recently been sent to Shanghai. She again kept a travelogue, this time with an eye toward its publication. Her report reveals how she was learning to see the world, both as big picture and in telling detail. The narrative shows her puzzling out her feelings about the competing ideologies surging throughout the continent, revealing her mixed emotions about communism and fascism. Her journal also sheds light on her social connections, and how she moved easily among people of disparate backgrounds, from wealthy diplomats to her fellow third-class passengers.

Three days after leaving London, Jaqueline arrived in Moscow, the western terminus of the Trans-Siberian railroad. Here she had entrée to the British Embassy, where Bertha Bramly's son-in-law, Aretas Akers-Douglas (Lord Chilston), was ambassador. In addition to hospitality in Moscow, Lord Chilston gave Jaqueline "a personal letter to show to stray Consuls." Her PEP friends also opened doors for her in Moscow. [3/4/34]

Jaqueline spent the next week on the eastbound Trans-Siberian express, where she shared a third-class carriage with three Russian men. Luckily, they got along famously; Jaqueline was studying Russian and they managed to communicate. She had hoped to steer clear of politics, but one night it was unavoidable. "Some form of central organization is clearly the only hope for states in the future," she politely agreed:

> I said, which was perfectly true, that I was highly sympathetic to them. As a matter of fact – provided one had had a "liberal education." I do think it would be more exciting to be a young Russian of say 20 than anyone else. There is everything to be done and some chance of doing it. [3/6/34]

Over the next several days, traveling through Manchukuo (the Japanese puppet state established in 1932) to Harbin and then via the southern branch of the Chinese Eastern Railway to Dairen, Jaqueline's fellow passengers were Mongol, Chinese, Japanese, and White Russian. These carriages were less comfortable and more crowded than the Russian train, and the ride grueling, although colorful.

In Shanghai, Jaqueline's observations of the English colony and regimental life were sharp and often funny. There were cocktail parties starting at noon, tiffen parties, tea parties, further cocktail parties, dinner parties, and outings to the cinema, sporting events, and nightclubs. Social life took place within the Westernized confines of the foreign settlement. After one long night of drinking and dancing she wrote: "it again seemed quite impossible that this could be a Chinese city half way round the world." [3/18/34] "Cosmopolitan gaiety & movement seems the keynote of Shanghai," she concluded. However, she had only a brief glimpse of the Chinese city, which was out of bounds for the regiment. [3/18–21/34]

Jaqueline had numerous family-related military ties to the English colony in Shanghai, and she was proud to identify with the Tyrwhitt military presence in China. Her brother Robert had been stationed at a naval base in Hong Kong. One night "a very distant Naval cousin" whose ship was then "in harbour," came to dinner. An invitation to a tea hosted by Cuthbert's battalion prompted Jaqueline to note that the barracks were the same "wooden huts that were run up for the emergency division summoned out by Cousin Reggie [Vice Admiral Sir Reginald Tyrwhitt] in 1927." [3/19–21/34] Sir Reginald Tyrwhitt, who had been the commander-in-chief in China, was principal naval aide-de-camp to the King; in July 1934 he was appointed an admiral of the fleet.

Signs of mounting international tensions dominated Jaqueline's tale of her journey back to London. She spent a day as a guest of the British consul in Dairen, where she noted "the atmosphere is totally Japanese." Perhaps predisposed to approve of another nation's imperial aspirations, she concluded (using the casually racist language of the time): "So far the Japs have done extraordinarily well I hear and though no one attempts to deny that their methods were entirely wrong in the first instance it is true to add that the parts of the country where they have as yet been able to establish control are now being quietly & efficiently governed & they have the general good will of the populace." [3/25–27/34]

On the westbound Trans-Siberian, Jaqueline began to notice the crowds were "more wretched looking." She learned from a Russian engineer in their carriage that "many people, especially the children, are very hungry." It dawned on her that "this train load has, generally speaking, nothing whatever to spend and the other [eastbound] had plenty." In Moscow, Jaqueline learned "it is authoritatively reputed that between 5 and 10 million died of starvation last spring … ." She had gained a more nuanced perspective on life in the Soviet Union: "At present things were by no means rosy but the future – always the future – had great possibilities." [3/31–4/1/34]

Jaqueline had mixed feelings towards two Germans who became her traveling companions in Dairen. She spoke no German but they both spoke some English. She initially viewed one of them as "a sentimental business man but honest and simple at heart," but it turned out that he was a "Party" leader. At the German border, he pinned a swastika on his chest and promptly began saluting every storm trooper on the train. Jaqueline had nearly a day to spend in Berlin, where she witnessed the menacing atmosphere of Hitler's regime. "Every fifth man was in uniform" and "hands shot out in all directions saluting at every moment." She passed by the old Reichstadt and "was amazed to find that there seemed to be most of it left even after all this fire trouble." Wandering the streets she observed:

> The windows of the picture shops were occupied by portraits of Hitler; those of the jewelers by swastikas and iron crosses in every size & metal; the bookshops stocked books on Hitler or the war & on physical training; the stationers were full of large scale maps of Germany; the tailors of uniforms; the shoe shops of hobnailed hiking boots and the hosiers of brown shirts collars & ties. Food & female fashions alone seemed to be immune from the infection. [4/6/34]

Traveling through Germany to Ghent, Jaqueline conversed with a man who was "violently against the present regime," although he wore a military badge. This man was "despairingly certain of war between Germany and France," a prospect she found both dismaying and hard to believe:

> I have now traveled through seven countries – Belgium, Germany, Poland, Russia, Manchukuo, Japan and China. All but Belgium & Poland are actually preparing for and talking of war and Poland seemed to be pretty thick with military men ... . I was never able to harbour the slightest doubt of Japan's aggressive intentions but I hate to begin to believe that the same is true of Germany. I can find very little that I have in common with this race, and the idea of the paralyzing influence of a widespread Teutonic domination is wholly horrible to me; but I am totally unable to believe that the public opinion in a country that has recently suffered so terribly from war can possibly want to repeat the performance. [4/6/34]

End of the League of Industry

The threat of fascism may have had something to do with the low attendance at the League's third annual conference in September 1934; there had been "allegations that the League of Industry is anti-trade union movement and a supporter of the Fascist movement," *The Times* had reported (League 1934: 14). The conference had considered the reports of its committees on planning for employment and agriculture; Jaqueline and Blaksley had contributed to the work on employment planning, which came out against the idea of a shorter workweek. The agricultural committee called for a protective tariff and noted: "A great deal more would be done for agriculture ... if the importance of the industry were made clear to the dwellers of towns" (Protection 1934: 11).

But by mid-1935 a combination of "better times and the resumption of party politics" convinced leaders of the League to disband (Mandle 1967: 39). The League came to an end in the summer in advance of a general election that would be held in November. Jaqueline lost her job, but in her years with the League she reported having gained "a thorough knowledge of the conditions prevailing in industry in the factories and the principal industrial personalities." [E.1] Plus, she now felt confident about public speaking and handling committees of workmen and employers.

Dartington Hall: Industry and Agriculture

Roskill probably used his PEP connections to the Elmhirsts to help Jaqueline line up her next job, at Dartington Hall. Since 1931, the Elmhirsts had been PEP's principle funder; they had hosted many retreats at their home on the estate, and Leonard chaired the agricultural research group. During the year Jaqueline worked at Dartington, from 1935 through 1936, learning about the association of rural industry and agriculture helped her integrate her concerns with socioeconomic reform and her earlier, enduring interest in landscape design. Arguably, it was the experience of being part of the multifaceted Dartington project that led her on the path to realize her vocation as a planner.

Dartington is situated in the bucolic Devon countryside, about 10 miles from the coast and close to wild moors and quaint villages, yet accessible to London by train. The Elmhirsts had acquired the derelict manor house and its 900 acres in 1925 with the aim of "rehabilitation in the broadest sense: physical reconstruction and redevelopment of all the resources of the estate in contemporary terms and scope for a full life for everyone connected to the Enterprise" (Bonham-Carter 1970: 19). As *The Times* reported in 1936: "The founders seek to reverse the drift of craftsmen to the towns ... and to re-establish the industries, with agriculture as the core, which can now be developed economically ... with the advent of electric power" (Dartington 1936: 15).

The Elmhirsts' experiment at Dartington was a cross-cultural hybrid, splicing the back to the land movement, a tradition with deep roots in England, with modernization along progressive lines primarily inspired by American practice. But the seed of the idea came from the work of the Indian poet and philosopher Rabindranath Tagore (1861–1941). In 1922 Leonard had helped Tagore create an Institute for Rural Reconstruction, Sriniketan, in Bengal, as part of what became Tagore's International University (Visva-Bharati) (Bonham-Carter 2007, Young 1982). Tagore encouraged Leonard, who became his personal secretary and disciple, to recreate Sriniketan in England. Dorothy had financed Leonard's work in India, before their marriage, and shared his commitment to Tagore's ideals of rural regeneration, progressive education, and the integration of the arts and sciences. When they married, they used her fortune to finance the Dartington enterprise, which Elmhirst referred to in 1929 as "our embryo Welwyn" (in Smiles 1998: 73).

By the mid-1930s Dartington was well known and a tourist attraction. An advertisement touting: "New Society – New Landscape" invited readers of *Rural Industries Quarterly Magazine* to tour the estate (Jeremiah 1998, 43). Much had been accomplished: major building projects completed; the coeducational boarding school opened; two scientific farms, horticulture and forestry projects established; new industries, including a sawmill, cider factory, textiles, a dairy, a commercial nursery and a building company launched; and an arts department created, with schools of theater and ballet.

The Elmhirsts then faced the problem of who would own the estate, which then comprised around 3,000 acres, and operate all of those undertakings. Their solution was a new technique of land management, actually a revival of positive aspects of feudal traditions embodied in the country house estate: "progressive management and social responsibility." In 1932, they conveyed ownership of the estate to the Dartington Hall Trust, charged with four main functions: to administer the Estate; promote education, both through the school and other educational services; encourage the arts; and undertake research, particularly in agricultural and woodland economics (Bonham-Carter 1970: 108).

Jaqueline was hired to help organize and run the new central administrative office that was created to coordinate these myriad activities. She worked in a modern building designed by the Swiss-born American architect William Lescaze. It was the latest of a series of buildings that Lescaze designed there, considered among the first important groups of modernist buildings in England (Gould 1998: 23). The modernist buildings represented the inter-related social-aesthetic ideals of the Dartington community, which the *Architectural Review* in April 1934, portrayed as "striving to build up a complete, purposive, fully conscious social organism ... because it is ... clear we cannot get out of the muddle in which *laissez-faire* has left us, unless we understand team work and have a new loyalty to a new order" (Heard 1934: 120–21).

Jaqueline's job situated her literally at the center of life at Dartington in the midst of a "brilliant period of the Estate's existence" (Bonham-Carter 1970: 132.) Like everyone she was encouraged to take advantage of all "facilities and means for enlarging the mind and sensibilities." It is easy to imagine her eagerly attending the Sunday evening meeting, where people would talk about their work, and Leonard Elmhirst would keep the community informed about his plans and ongoing projects. Resident and visiting teachers offered "practically every kind of intellectual and artistic performance." The high proportion of foreigners among the artists employed at Dartington 1934–1940 made those "six brilliant exotic years" (Bohnham-Carter 1970: 135). Dartington was especially known as a meeting place for "kindred spirits" of East and West. The arts definitely enriched Jaqueline's life at Dartington, and she made several lasting friendships.

Dartington was one of the first places in Britain to give asylum to refugees fleeing from Nazi Germany and later, the Spanish Civil War. The Elmhirsts supported several former faculty and students immigrating to Britain when the Nazis closed the Bauhaus, the modernist design school that Walter Gropius founded in Dessau,

Germany, in 1919. They helped Gropius form a partnership with architect Max Fry, who was both a member of PEP and active in the Modern Architectural Research (MARS) group, the British branch of *Congrès Internationaux d'Architecture Moderne* (CIAM), the group of avant-garde architects that Gropius helped form in 1928 (Bonham-Carter 1970: 135). Jaqueline would have taken great interest in the work of Gropius and Fry at Dartington while she was there: the conversion of an old barn to a theater.

Dartington was also famous for its gardens. Jaqueline surely relished this aspect of her surroundings, and she cultivated a friendship with the head of the gardening department. At that time planting of the courtyard gardens and grounds around Dartington Hall was underway; overseen by a noted American landscape architect.

Turning Point

Surely the experience of living at Dartington was in itself a significant factor that led Jaqueline to decide to study regional and town planning. Tyrwhitt later recalled that it was in 1936 that she came across the prospectus for the new School of Planning and Research for National Development (SPRND); "at once I decided that this expressed my aim in life." [T.3] SPRND was an offshoot of AA, which had decided to start a planning school in 1934, shortly after passage of the Town and Country Planning Act. Most of the School's Advisory Board, including the founding principal, E.A.A. Rowse, was followers of the pioneering biologist and town planner Patrick Geddes. It is interesting to note that Geddes and his son Arthur, a geographer, had a special connection to Leonard Elmhirst and Dartington through their work with Tagore in India. Both Patrick and Arthur Geddes corresponded with Leonard concerning agriculture and town planning, among other topics. Arthur translated Tagore's songs; they worked on this project at Dartington when Tagore visited in 1930. Jaqueline might easily have encountered Geddes's ideas at Dartington before she studied at SPRND. She recalled: "It was Patrick Geddes's *Cities in Evolution* that first got me started on an interest in town planning." [T.4] That book was out of print but there was probably a copy in the Dartington Library.

Jaqueline seemingly spent a happy and rewarding year at Dartington. She maintained contact with the community for years, returning to visit friends, sending her goddaughter to the school, and staying active in an alumni group. Yet, she left her job there in early 1937 "as she considered the aims of the organization were becoming too closely connected with party politics." [E.1] Leonard Elmhirst and many others at Dartington were associated with socialist thinkers and ideas, but reportedly there was no dogma to which members of the community were supposed to adhere (Bonham-Carter 1970). Jaqueline's discomfort may be related to the campaign leading up to Leonard's election to the Devon County Council in 1937. As Devon was a socially very conservative county, it was a hard-fought election. It is easy to imagine that Jaqueline felt awkward in those circumstances given her former Conservative Party activism. Her sensitivity could have been heightened by the tensions surrounding the abdication crisis, culminating in December 1936, when King Edward VIII renounced the throne in order to marry the American divorcee Wallis Simpson, who was suspected of having ties to Nazi Germany.

Whatever the reasons for Jaqueline's departure from Dartington, her next move is perplexing. Having unexpectedly come into some money, she "went to Berlin for nine months to study Town Planning, particularly Land Settlement." [E.1] Her decision to study in Berlin, the capital of the Third Reich, was a short detour on the path of her intellectual and political transformations, but it had consequences.

Chapter 3
Studying Regional and Town Planning

To Germany

Jaqueline spent the first nine months of 1937 studying town planning and land settlement at the Technische Hochschule in Berlin (TH–Berlin). She later stated that she went to Germany "partly to see what had happened to the earlier town-planning schemes under Hitler." [E.1] But her further explanation that she also went "partly to experience life under a totalitarian regime" suggests there was more going on than she was willing to acknowledge. In the absence of any first-hand account, it is impossible to know what Jaqueline was thinking before, during, and immediately after her time in Germany. Given her penchant for keeping travel diaries the absence of any from this trip is noteworthy. But her subsequent writing and work indicate that it was a vitally important experience that helped focus the intellectual transformation she was undergoing. To understand what may have attracted Jaqueline to Berlin in 1937, it is useful to consider what she probably knew of the "earlier town planning schemes" she referred to and contemporary accounts in English about the achievements of National Socialist planners.

English Literature on German Town Planning

As noted previously, Jaqueline credited reading *Cities in Evolution* by Patrick Geddes as having sparked her interest in town planning. That book singles out German town planning for praise. Geddes (1915: 175, 220) particularly admired the way in which modern German urban development preserved and revitalized "the antique spirit of her great free cities of the Middle Age." He thought German town planning at its best demonstrated:

> a growing association of civic and social action with architectural and artistic effort also in unison, and these to a degree lacking in our British towns, or lapsed even where in past time it has existed … . German cities are actively entering upon this new advance of city life and creative art together: and now also is our own opportunity. We have to live in towns: and, on the whole, with all respect to Garden Cities and Garden Suburbs, we have to make the best we can of the existing ones.

One of the young planners who admired and worked with Geddes was George Pepler (1902–1959), who was instrumental during the 1920s in monitoring German innovations in urban planning on behalf of the Ministry of Health and as a leader of the International Federation of Housing and Town Planning (IFHTP), which he served as President from 1935 to 1938. Another follower of Geddes, architect Sir Raymond Unwin, was influenced by German town planning in his proposal for a green belt around outer London, adopted 1935. Unwin's design for Letchworth, the first garden city, had earlier inspired leaders of the German garden city movement, who were among those commissioned to design the large housing estates (*siedlungen*) built in many German cities by local governments during the short-lived socialist Weimar Republic (Ward 2010). These estates attracted British attention; they epitomized the new socially conscious architecture called for by Gropius in founding the Bauhaus School in 1919 (Lane 1968).

Jaqueline was aware of Gropius's ideas not only through Dartington and PEP circles, but also because her mother's cousin, architectural critic Morton Shand, had helped Gropius emigrate to Britain in 1934, and translated his *New Architecture and the Bauhaus*, which was published in London in 1935. In this book, Gropius addressed the relationship of the New Architecture to the task of Town and National Planning, which began with the distillation of standardized dwelling types suitable for mass production as the basic unit of urban design. Jaqueline could easily connect Gropius's emphasis on standardization to her father's work with schools in the beginning of the century. Jaqueline may have also read Lewis Mumford's *Technics & Civilization* (1934), and Catherine Bauer's *Modern Housing* (1935), both of which praised the *siedlungen* as models of progressive planning practice.

National Socialist Regional and Town Planning

By 1937 the Nazis had implemented many of the urban renewal programs that had been planned during the Weimar Republic but not realized due to economic and political turmoil. But, under the Nazis, new housing was built to a suitably *Völkisch* vernacular style rather than modernist design. "The ideology of *Blut und Boden* (Blood and Soil) called for housing produced with traditional arts and crafts techniques and situated in greenery, thereby enabling Germans to become "attached to the soil" (Diefendorf 1993: 158, 113).

When Jaqueline started at TH–Berlin, the faculty who survived the Nazi purges would all have supported the party line. Jaqueline would have studied regional planning with Gottfried Feder (1883–1941), a strong believer in Nazi anti-urban, *Blut und Boden* doctrine. As Reich Commissioner of Settlement in 1934, Feder announced his aim to be "the dissolution of the metropolis," and called for "the reincorporation of the metropolitan populations into the rhythm of the German landscape" (Schenk and Bromley 2003: 113–14). While his ideas were too radical for practice, Feder used his time and his students—possibly including Jaqueline—at TH–Berlin to research a National Socialist theory of organic town planning, published as *Die neue Stadt* (1939).

Feder built on the English garden city model in proposing the redistribution of the German population to small cities of a maximum of 20,000 inhabitants each (Schenk and Bromley 1965). His ideal Nazi urban form was oval or circular, adjusted to fit local topography, with a civic core, a radial street pattern, and

Völkisch style neighborhoods, each centered on a school and other community facilities. He envisioned that each new town, preferably situated in a rural setting, would function as the service center for its "influence circle" of agricultural villages, based on geographer Walter Christaller's central place theory. This nodal pattern not only preserved historical settlement patterns (*Heimat*) and achieved economic efficiencies, it also reflected a strategy to "Germanize" the landscape and population in occupied Eastern lands (Mantziaras 2003: 161).

Lebensraum and Regionalism

At TH–Berlin, Jaqueline would have been introduced to the Nazi government's relatively new program of national and regional planning and development. There was great interest in this topic in many countries, including Britain, France, and the US, which were then concerned with regional reorganization. A National Planning Board responsible for the coordinated planning of the Reich was created in 1935. The National Board oversaw 23 newly established Planning Regions, which coincided with political units. A Regional Planning Federation supervised production of a comprehensive regional plan. A Regional Planner did the actual work, based on "an elaborate scientific investigation of the conditions and problems peculiar to the region." A National Board for Areal Research coordinated the necessary research to be conducted by interdisciplinary university-based working groups (Dickinson 1938: 624).

Contrary to *Blut und Boden* doctrine the new planning regions corresponded with existing economic regions, and Germany's metropolitan centers were the heart of its major functional regions. British geographer Robert Dickinson (1938–1939: 30), who conducted research in Germany in 1936–1937—and whose interest in regionalism was influenced by Geddes—highlighted the tension between a planning system based on functional regions and the Nazi's anti-urban ideological mandate, which he explained in terms of the concept of *Geopolitik*. Geography was seen as the basis of regional planning that, as stated by the director of the National Board for Areal Research in 1936 (in Dickinson 1938–1939: 31–2)—in a twist on Geddes's credo: "place, work, and folk ... as elements of a single process—that of healthy life for the community and the individual" (1915: 198)—should take "due regard of area, folk, and race as formative forces in the evolution of the region as an organic entity." *Geopolitik* warped teaching and research in planning, utilizing "scientific methods" to justify an ideological, racist agenda.

By the time Jaqueline arrived at TH–Berlin, increasing militarism was leading the Third Reich to de-emphasize its domestic policy of urban decentralization. A new emphasis on military industries refocused attention on major cities and ideas for their redesign. This new emphasis would have been underscored at TH–Berlin since, in January 1937, architect Albert Speer, a prominent graduate and teacher, was charged with the redesign of Berlin as a monument to Nazi power and ideals (Mullin 1982).

School of Planning and Research for National Development (SPRND)

Jaqueline returned to London in October 1937 along with a friend from TH-Berlin, "Horsi" Raffloer. Both women enrolled in the two-year post-graduate diploma course at SPRND, then still in its start-up phase. As noted previously, the principal of the school, EAA Rowse (1896–1982), along with several members of the school's Advisory Board, notably Pepler and Unwin, were great admirers of Geddes, whose ideas provided the conceptual framework for SPRND's curriculum, specifically: that a statutory plan must be preceded by both a regional and civic survey. In contrast with existing programs that were civic design oriented and accepted only architects, engineers, and surveyors, SPRND also admitted graduates who had studied subjects such as sociology, geography, and economics as such skills were needed for this comprehensive approach to planning. [T.5]

In addition to its broader curriculum, SPRND aimed "to widen the [spatial] field of its teaching on a National scale" based on the conviction that "the Planner capable of coping with national problems will, before long, be called into existence by the needs of the times." The school's prospectus went on to address concerns about the compatibility of planning and freedom in light of the rise of fascism on the continent and communism in the Soviet Union (APRR 1948b: 1):

> There is found in the School of Planning and Research for National Development no inference that the regimented nation is the ideal political body. The School seeks to create ... a corps of trained men, possessing the necessary breadth of outlook and technical knowledge, whose collaboration with and ultimate succession to those who now perform similar tasks will ... ensure happy and ordered development in the place of the chaos which the nineteenth century has left us.

When Jaqueline began at SPRND the school was still closely connected to AA, where Rowse had also served as Principal since December 1935. Moreover, the two schools were neighbors on Bedford Square. Jaqueline recalled that when she arrived "the School had parted from the AA on the official level but on the student level there was still a lot of to-ing and fro-ing." [T.2] In January 1938, in his capacity as AA Principal, Rowse even invited Jaqueline to serve on a panel of Honorary Visiting Critics to judge student presentations. [T.6] AA was then in the midst of turmoil, triggered largely by Rowse's efforts to reform architectural education there. Rowse's re-organization of instruction caused a controversy that culminated in his forced resignation in May 1938.

The reforms that Rowse had begun to introduce at AA, in the 1936 spring term, were part of a larger effort to redirect the school away from its traditional Beaux Arts orientation and toward one aligned with modernist principles such as Gropius had instituted at the Bauhaus. The most active AA students and younger faculty supported Rowse's approach, notably the team that worked on the famous "Tomorrow Town" project (Darling 2007, Crinson and Lubbock 1994, Ashton

2000). Those were the AA students and faculty who engaged in the "to–ing and fro–ing" with SPRND, and Rowse pointed to "Tomorrow Town" to promote the work of SPRND (1939: 169).

One reason why Rowse could not point with equal pride to a project produced by SPRND students in 1939 is because the school never quite coalesced before the war perhaps in part because Rowse was distracted by the tumult at AA, at least until June 1938. Jaqueline recalled: "We were plunged into a maelstrom of seemingly haphazard lectures from all sorts of people ... . As one entered [the school], after having done a day's work, one could never be sure what the evening might include." [T.3] One member of the faculty, geographer E. C. Willatts, who taught "Economics as Related to Planning" ("better described as aspects of geography related to planning") confirmed: "About 50 lecturers, many of them very distinguished, were recruited to give short courses of lectures. Perhaps there were too many teachers, but at least future planners were introduced to the importance of many problems and ideas" (1987: 103).

It was hard for SPRND to offer a more organized sequence of lectures since the school had no source of funding other than student fees. Rowse sustained its activities "more or less single handedly" (Meller 1990: 323). "No payment was made to lecturers at all, and though the school was able to tap the best brains of the country on the subjects Rowse had decided we should be exposed to, it was not easy to persuade them to turn up in any logical order," Jaqueline explained: [T.3]

> Certain courses were regular, but the rest was a bouillon. This suited me, after I'd got over the first shock; ... Gradually we began to perceive the common threads running through the divergent talks – though I feel sure that some of the younger ones who had come straight from school and AA were as much in a fog at the end of the course as at the beginning.

Moreover, in stark contrast to the teamwork Rowse encouraged at AA, students at SPRND were pretty much left to their own devices. "Our studio work was done at home over the week-ends and we discussed this with Rowse and [his assistant] after the evening lectures were over—when we were all pretty tired," Jaqueline reported: [T.3]

> Criticism consisted almost exclusively of questions ... . As far as I remember we were very seldom given detailed instructions about how to find out our answers. We were given introduction to individuals or organizations when we asked for them ... . The pity was that there was very little close contact between the students themselves; we only met in the evenings as part of a common audience.

Once Rowse could devote his full attention to the school he prepared a "further revision to the curriculum, giving fuller expression to [the] ideal" of teamwork (1939: 168), which he submitted to the school's Advisory Board in early 1939. Jaqueline was nearly done with her coursework by then.

Nevertheless, Jaqueline was satisfied with her education: "By the time I had finished the course I felt I really had been made to grasp how to go about a job, and that I knew a good deal about the whereabouts of source material." She was fortunate to have had access to the large network of prominent lecturers and "friends" of SPRND. Jaqueline singled out Pepler—then the Chief Town Planning Inspector to the Ministry of Health (1919–41), responsible for approving statutory plans prepared by local authorities—for his "warm and friendly help." [T.3] Two lecturers stood out for Jaqueline: Professor Eva Taylor—who was then Chair of Geography at Birkbeck College, University of London—on Geography; and Rowse on History. While studying at SPRND Jaqueline followed Rowse and Taylor in becoming "an ardent disciple" of Patrick Geddes. [T.7]

Regional Development: Location of Industry

One reason Jaqueline was able to get so much out of the School of Planning evening course despite its chaotic nature was because the work she did during the day reinforced those lessons. In the first year (1937–1938) she divided her time working with two stalwarts in her life throughout the thirties: Blaksley and Roskill. Blaksley came to Jaqueline's aid when she needed a job, just as he had done in 1931. He arranged for his then employer, the National Institute of Industrial Psychology (NIIP), to hire Jaqueline to work part-time for a year on his study: "Leisure Pursuits Outside the Family Circle."

Her work, which began in December 1937, involved reporting on conditions in one of the three cases studied: a Country Town, a Satellite Town, and a Rural District and Holiday Camp (Bevington et al. 1939). This represented a notable shift from the activist roles she and Blaksley played when they worked as organizers for the League of Industry. Under the auspices of NIIP, Blaksley and Tyrwhitt now applied "impartial" scientific methods to study how to improve worker productivity, by determining which obstacles prevented workers from unwinding on their time off. This was exactly the type of scientific study of behavioral patterns that Rowse encouraged at SPRND, to improve regional planning, and civic design.

In her spare time, Jaqueline worked on the PEP Industries Group's report on the Location of Industry. She likely learned as much, if not more, about techniques of fact-finding and analysis from Roskill, the Industry Group's secretary, as from the SPNRD lecturers. The PEP Executive engaged Roskill's management consultant firm to prepare the factual material for this report, and suggest conclusions that followed from the facts for discussion by the Industries Group. This group invited "distinguished leaders from the industry under discussion to join it either as members or in a consultative capacity" (Roskill 1981: 56). As noted previously, Jaqueline had established an informal connection to the Industries Group from its inception, via her friendship with Roskill and her work for the League of Industry. Given those existing relationships, her subsequent work at Dartington, first- hand knowledge of current German planning, and her connections to leaders in the field through SPRND, she was well qualified to contribute to the Location of Industry study. But the formal invitation to Jaqueline to join them as a consultant, the term PEP used for volunteers, is particularly noteworthy, as she became one of only five women to take an active part in PEP activities in the next 50 years (Nicholson 1981: 39). However, her contribution was cloaked in the anonymity that was a hallmark of PEP's approach. An official description of the final report—the "result of four years' work by an independent group of professional men, mainly economists and civil servants" (McCallum 1945: 292)—erased her gender as well.

The preliminary findings of the Location of Industry study had been published as a PEP broadsheet (no. 87) in December 1936. This was followed by appointment of the Barlow Commission on the Industrial Population, headed by Sir Montague Barlow, in July 1937. Jaqueline was involved as members stayed in constant touch with Sir Barlow and kept him apprised of their work. The Industries Group report, "Location of Industry in Great Britain," published in March 1939, anticipated and according to Roskill (1981: 64): "might with justice claim to have had a significant influence on" the Barlow Commission's Report, notably the recommendation for an Industrial Development Commission to guide the location of industry taking into account the concentration of growth in a core and periphery of industrial stagnation or decline. This report—"one of the most influential single forces in shaping Britain's post-war welfare state"—demonstrated that the problem of concentrated industry and the problem of urban sprawl were inter-related (Hall 1981: 147).

Based on her high level of commitment to PEP, by 1938 Jaqueline had shifted definitively away from identifying with Blaksley's right-wing Tory ideal and was proud to be associated with the "middle way" consensus that Roskill, via PEP, was a leading force in shaping. PEP was "not Conservative either in the party political or in the traditionalist sense. It questioned the Status Quo rather than supporting it," Roskill (1981: 65) attested. But he added: "we were, after all, a Group which believed in private enterprise. The furthest the Industries Group would have gone would have been to accept that in some cases, for example where monopoly appears to be in the public interest ... public ownership should follow." Roskill described the outlook of people who joined the Industries Group by defining what they were not: "people who accept the ethos and customs of those in authority and climb the established ladders of advancement in whatever occupation they follow. An organization which sets out to question current assumptions, to assemble and study the facts and see in what direction they point is unlikely to find immediate favour with those responsible for keeping the machine running."

Garden Cities

During Jaqueline's second term at SPRND (1938–1939) she was employed part-time as a Research Assistant for the Garden Cities and Town Planning Association (GCTPA); this job was indirectly connected to the work of the

Barlow Commission. FJ Osborne, Honorary Secretary of GCTPA, had revived that nearly defunct organization—which he had co-founded with Ebenezer Howard, among others, during World War I—primarily to influence the outcome of the Barlow Commission's efforts (Buder 1990). Towards that end he created the Welwyn Garden City Research Committee in 1937. Jaqueline's job was to coordinate grant-funded research on the problems affecting the conditions of the working population at Welwyn Garden City, where she was also obligated to live. Jaqueline rented a small, semi-detached cottage at Welwyn which she shared with Horsi Raffloer. She conducted the survey there under the guidance of staff from the newly established National Institute of Economic and Social Research at University College London. [T.8] Rowse agreed to accept this survey in lieu of the research requirement to complete her course work.

In late July 1939, Jaqueline and Horsi both passed their final SPRND exams, although Jaqueline still had to complete her Welwyn survey in order to earn her diploma and qualify for membership in the Town Planning Institute (TPI). Jaqueline was proud to be invited to give a lecture at the next TPI Summer School, scheduled for September 1. However, Germany invaded Poland that day and the lecture was canceled. Horsi had returned to Berlin by then, leaving Jaqueline alone in Welwyn Garden City. She invited a refugee German Jewish family, the Hammerschlags, to live there. Jaqueline knew their daughter Steffi from Dartington, where Steffi had been a trainee in the Nursery School. The Hammerschlags, who had only escaped from Germany the previous week, could pay rent if and when Jaqueline was sent away to serve in the Women's Land Army, which she had joined soon after it was formed in June. [9/3/39] Jaqueline ended up moving out of her now crowded house to stay with a friend.

Jaqueline included this information in a letter she wrote to Leonard Elmhirst in late September. She had heard from another Dartington friend, the musician Nannie Jamieson, that Elmhirst, who was now Chairman of PEP (1939–1953), had "an extremely interesting scheme afoot." She let him know: "If there is any work you think I could do to further your schemes, and that would justify me in asking to be released from the Land Army, you can count on me." [T.8] This request indicates that the evolution of her ideas was nearly complete.

SPRND closed its doors with the outbreak of war; the curricular changes Rowse had proposed weren't ever implemented. Most of the staff and students joined the Army, and the Advisory Board ceased to meet. Jaqueline wasn't only among the first graduates of SDRNP; she was also one of the last, at least in its original incarnation.

Chapter 4
Serving on the Land

Planting Seeds

In September 1939, Jaqueline found herself in an uneasy state of transition and profound dislocation. Her father, now a widower, had moved into a small mews house, feeling a sense of duty to remain in London. Her sister, Edrica, had moved with her two young children to northern Wales, near her husband's family. Her brother Cuthbert had been transferred from his post of military attaché in Peking (Beijing) to a post in Singapore; his wife, Delia Scott, an American socialite whom he married in Peking in March 1936, soon joined him here. Moreover, Jaqueline had received a month's notice from Osborn.

She was called up for service in the Women's Land Army (WLA) and told be ready to report on October 1, 1939. In choosing to serve the war effort on the land, Jaqueline was reaffirming her roots in the women's garden movement, which had given rise to WLA to aid in food production during World War I (King 1999, Sackville-West 1944). Unlike the other women's auxiliary forces, WLA was a civil rather than a military service; WLA volunteers were employed by private enterprises. Jaqueline was hoping to do market garden work. [9/22/39] While waiting for a suitable WLA position to open, she enjoyed a restful time in the Sussex countryside, writing up her Welwyn Survey in order to receive her diploma and her certificate of membership in the TPI. [T.9] Fortunately, Osborn extended Jaqueline's employment for another month so she could finish the report. She stayed at the house of a friend in the garden city, where the GCTPA office was located, and in exchange helped take care of her friend's young children—a role she played frequently during the next decade.

As the weeks passed, however, it appeared as if a job with the Land Army might not pan out. Tyrwhitt could not afford to be out of work and could not stay in Welwyn. She reached out to Rowse, who was now conducting school business from his pig farm in Sussex, for help in finding a job on the land. [G.1] Rowse had his own problems with a disorganized Land Army, though. "I am not in the least surprised that they have no need of one with the specialized knowledge and efficiency you possess," he replied. [T.10]

Tyrwhitt's old mentor and friend, Chrystabel Procter, who was now the Garden Steward at Girton College, Cambridge University, saved the day. When Procter's foreman fell ill, she hired Jaqueline to take over in the interim. This was a big job, since the college grew nearly all of the fruit and vegetables it consumed. (Jaqueline knew the gardens well as she had helped lay them out in 1933.) "I like the life – and it's a relief to be settled, even as a stop-gap!" Tyrwhitt wrote to

Francis Davies, Rowse's secretary, in early November; Tyrwhitt asked Davies to remind Rowse to "fix things about that 'research work'" so she could "get the TPI settled." [G.2]

Jaqueline had already received Associate Membership in the Institute of Landscape Architects (ILA), in May 1939, no doubt with Procter's encouragement. Both were members of the Women's Farm and Garden Association (WFGA), which promoted the profession. Jaqueline wasted no time in cultivating this new professional network. Architect Sir Geoffrey Jellicoe (1900–1996), who was a cofounder of ILA in 1929—and who succeeded Rowse as Director and Principal of the AA—recruited Jaqueline to join ILA's newly formed Committee on Allotments. Jaqueline also befriended Brenda Colvin (1897–1981), another ILA founding member, with whom she shared a love of trees. Likely encouraged by Procter and her ILA colleagues, Jaqueline decided, given the dearth of Land Army openings on farms, to seek a position in forestry; specifically to become a "timber measurer," a job particularly suitable for women with some architectural training. [T.11]

Women's Land Army: Forestry Service

Tyrwhitt had a longstanding interest in Forestry through her uncle, R.E. Marsden, who had served as a Research Officer under the Forestry Commission in India. Jaqueline asked for his help when she heard that WLA headquarters was about to begin a Forestry Course with a few spots for special recruits as timber measurers. The instructor was an old acquaintance of her uncle's, and happy to include her. [T.12]

Jaqueline's Timber Measuring course began in January 1940, at the Forestry School in the Forest of Dean. From there she was posted to Linwood in the New Forest, in southern England. "I measure the intake and output of two small forest timber mills, arrange about dispatch and so on, and–in between whiles–work in the forest itself: 'browst burning,' 'brashing,' 'dressing out,' and what not." She wrote Rowse in March describing an idyllic life:

> I have found digs in a small dairy farm, and in the evenings I milk cows, ride over the hills to fetch them on a forest pony, do gardening and so on. I have also developed a small fox terrier of a hardy nature [Jill] whom I am busy training as an outdoor dog) ... . My work necessitates walking 6 or 7 miles a day through the forest – observing as one goes–and my actual job is with and for forest craftsmen–"real" people. The house I live in is 6 miles from a village and 8 miles from a station. We pump our own water, deal with our own sewage, live off our own land – 18th century except that we have oil lamps. I believe that a few months–perhaps even a year or two–of this life will teach me any things that are really important for regional planning–and I am not anxious to interrupt the course. [G.3]

Call to Service: Planning for Post-war Reconstruction

Jaqueline provided those details to explain why she felt "of two minds" regarding Rowse's offer of a job working on a plan for the redevelopment of the South Wales coalfields. Prior to this offer, Rowse had tried to convince her to join with fellow former students who had worked together on a project for East Anglia, to form a professional collaborative. His model was the Architects' Cooperative Partnership (ACP), which had been formed in early 1939 by the team of recent AA graduates who worked on the Tomorrow Town project. He wanted to present both the East Anglia and Tomorrow Town projects in an exhibit he was organizing, but not as the work of students. [G.4, JT.13]

Subsequently, Rowse had decided to take the lead in forming the collaborative in order to undertake the South Wales scheme himself, under the auspices of the School of Planning. With potential funding from Lord James Forrester (1919–1960), a philanthropic industrialist, Rowse hoped to hire two people to work on the project for about a year. He had proposed Jaqueline "as the senior in charge of the development of drawings, reports, etc.," with Peter Cocke, a member of ACP, "as an assistant collaborator." The rest was sketchy. Rowse explained: "A group of us are to advise on the work, with a panel of experts and the Executive Committee of the School co-operating." [G.5]

At the same time, Jaqueline heard from Frances Davies about Rowse's plans to reorganize the School as an international correspondence course. [T.14] The time was ripe for such an undertaking. Civilian groups concerned with adult education had convened a Central Advisory Council for Education in H.M. Forces, which had worked with regional committees at each university and college to prepare preliminary schemes for a range of educational opportunities for the troops, including correspondence courses in professional and technical subjects. The War Office had authorized its own committee to devise a scheme of Army education in wartime (Hawkins and Brimble, 1947).

But Rowse's plans for the reorganized school were subordinated to his efforts to get the South Wales scheme going. He was likely inspired by the formation in February 1940 of "The 1940 Council to Promote the Planning of Social Environment." The Council, an all-party organization under the chairmanship of Lord Balfour of Burleigh (a member of the SPRND Advisory Board), sought to raise private funds to support local efforts to undertake planning through research groups and other means, based on the principles set out in the Barlow Report (Hardy 1991).

Jaqueline assured Rowse that she would get her East Anglia maps ready for his exhibit; and she would think about a "group name"—"something that stressed the 'regional' rather than the 'town' planning aspect." But her missing Diploma remained an issue. Although she still intended to send him her final Welwyn Survey report as her "research work" for the Diploma, Jaqueline suggested she would be willing to take the job if it would get her TPI membership. "I know one should ignore and despise labels – but they have their uses, especially for women!" [G.6]

Rowse responded with flattery and appeals to loyalty: "I have cracked you up heavens high, so you must live up to your reputation and not let me down." However, he avoided the issue of the Diploma and was dismissive of Jaqueline's concern for credentials, saying: "to all intents and purposes you are labeled already. I am thinking more of your life's work than any alphabetical strings coming after your name." [G.7] In mid-April Rowse wrote again: "A world problem has to be solved when dealing with the difficulties of South Wales. Can we, who have been specially trained for this task, not help by attempting to find a solution?" The junior position had been eagerly accepted. Yet Rowse now cautioned: "Do not go and throw in your job at the moment, however, should you decide to join us; wait till I am absolutely certain that things are above board." [G.8]

In the interim, Jaqueline's situation changed. On April 13 her landlady gave her a week's notice. "I think I get on her nerves" she wrote in her diary. Jaqueline was more receptive to Rowse's proposal under those circumstances. She immediately responded, asking for details about Lord Forrester's background, the other financial backers, advisory board members, the motives behind the scheme, implementation issues, the scope of the study, and logistical aspects. [G.9]

Rowse responded promptly and defensively. "You should not have asked" about motives, he scolded. "Some of us are not acceptable in the Forces as yet – in the meantime we ought to do something of service to those who may come back to a very unhappy world, if the present opportunity is missed." He finessed the issue of implementation: "I think that if we are successful in rousing South Wales itself to action, we can make a sufficiently healthy row to hot up the atmosphere at Westminster to an uncomfortable temperature." He also chided her for questioning the scope of the study: "one of the reasons why I hoped that you would come in, was that you had already very fully grasped the technique of the definition of provincial and regional boundaries and could, further, go about such work as a preliminary step without a lot of delay and explanation." [G.10]

Rowse's admiring description of Lord Forrester—who would provide the primary funding for the project through his family's firm, Enfield Rolling Mills— was Jaqueline's first introduction to a man to whom she soon became devoted:

> Lord Forrester is ... very quiet and purposeful, still young with a quite useful double background of industrial experience ... and the credit of the sounder work put in on the Brynmawr experiment. He was responsible for the organization of the factories and the subsistence farms and, by all accounts, did it exceedingly well. No politics, as far as I can gather, whatever. Rather pro T.U. [trade union] than anti. [G.10]

This information, combined with her unsettled living situation, convinced Jaqueline to accept Rowse's offer, although she still hoped it would fall through. Rowse's stern tone had revealed that he was more involved than she had realized. "I am ready to do all I can to help ... . I admit that some of my questions were too impertinent to put to someone personally connected with the job ... I've been pretty badly burnt ... by plunging into schemes that I'd accepted at their face value." She noted that her skepticism was justified, based on the "unfortunate" history of plans for South Wales. Lord Portal, whose industrial survey in 1934 led to the designation of Special Areas, "had the personal backing of Edward VIII;" Malcolm Stewart, the first Commissioner for the Special Areas, had "the whole of the Quaker & banking world behind him;" and Peter Scott's Brynmawr Experiment "never stirred the waters much outside the Eastern Valleys." [G.11]

On May 3, Rowse reminded Jaqueline "not to quit her job yet." [G.12] A week later the German army invaded Holland, Belgium, Luxembourg and France, and Winston Churchill replaced Neville Chamberlain as Prime Minister.

Suspicions

The intensification of war triggered new concerns within Britain about the domestic fascist movement, which had enjoyed support among the British aristocracy. In May 1940, the government arrested hundreds of fascists believed to represent "a fifth column working for German invasion that now seemed imminent" (Pugh 2005: 3). In late May, Britain suffered heavy losses in the evacuation of troops from Dunkirk. In this atmosphere of mounting fear and confusion, Jaqueline came under suspicion as a Nazi sympathizer.

Jaqueline reported in her diary that on May 31, 1940, Colonel Ogilvie, a retired military man, came to the mill and made inquiries about her. Ogilvie asked her employer about her hours, did she have any sweethearts, and did he know that she had recently returned from Germany? A few days later a police sergeant interrogated her about how she had acquired British Nationality, and about how many relations she had in the German army. The Colonel complained to the Chief Constable about her. She confronted the Colonel but he "would give not ground for his suspicions … . What it is to live in a village!" Jaqueline's notoriety caught the attention of one of the local gentry, Major Jarvis. Upon learning that she knew the Bramlys, he made clear (in public) that she was always welcome at his home. "Little snob!" she complained, recalling the months she had been ignored. While her connections to the elite suitably impressed the local authorities, those ties may have also provoked those who were suspicious of her presence there.

In her (available) diaries and correspondence, Jaqueline does not reflect on how either her past politics—her membership to the British Fascists, or her sojourn in Nazi Germany—might have contributed to her troubles. The experience left her shaken. In a letter to Rowse on June 14, the day Paris fell to the Germans, Jaqueline made light of the affair and offered a theory: "I've had the honour of being interrogated as a spy because I spoke German & possessed plans of East Anglia." [G.14] Days later the Germans extended their British bombing targets to include major coastal towns. The first major air-raid over Southampton, about 20 miles away from Linwood, where Jaqueline lived and worked, came on June 19. On June 30 the German army occupied the Channel Islands of Guernsey and Jersey.

The next day the police interrogated Jaqueline again. She vented her frustration in a letter to Rowse, elaborating her theory that her first landlady became convinced she was harboring a German spy after seeing the East Anglia plans. Rowse had relocated School of Planning operations from his farm to the Building Centre in London, where FR Yerbury (a cofounder of the school and Advisory Board member) had opened an atelier for displaced architects. He promptly responded on July 5 with promising news. He, Yerbury, and Forrester had agreed on a more doable approach: they wanted to hire Jaqueline to prepare a model plan for the eastern valley of the coal-fields area that could be applicable elsewhere in the region. She was to start as soon as convenient. After she "had absorbed the preliminaries of the problem," she was to go to South Wales "to make contact with those who will organize the Welsh side of our team work." Rowse was unsure about his own role. He was eager to join the fighting, but might not be called up for some time, if at all. "We might have time to work together on the development of the outlines of a general plan for South Wales, while you are going ahead on the small problem." [G.14, 15]

With this offer—and smarting from the spy affair—Jaqueline wrote to the Forestry Commission requesting to be released from her position at the mill. [T.15] However, she immediately had second thoughts: "Can't think why I make myself go when I am so comfortable and content here." [7/11/40] Rowse sent an outline approach, leaving the details up to her. He also advised her that he was stepping up his effort to enlist. [G.16]

Jaqueline responded suggesting additional advisors, and asking that PEP be among their collaborators. "I have already asked & obtained permission to use the relevant material collected by their Regional Research Group–my E. Anglia boundary map was largely based on this," she informed Rowse. [G.17]

As Jaqueline's preparations advanced her ambivalence about the new job increased. She was happy in her new lodgings and at the mill: "I watched the sweeping skies & heard the swish of the forest as a breeze came up behind me—& wondered what it was that drove me to take on an arduous & ungrateful task & leave this. That self torture that drives one from work that is easy & pleasant to work that is always just too hard to tackle." On her last day she treated the men to beer and cake. [7/31/40]

Turnabout

Before returning to London Jaqueline spent a few days at the home of friends. One evening the butler announced that there were two men to see her at the back door: her former employer and a co-worker. [8/2–4/40] Her employer had been asked to expand his output at the mill and start another one, to fill the urgent need for timber. He felt he could not do this on his own and threatened to resign, which meant the mill would close, unless Jaqueline returned. In addition to her duties as a Timber Measurer, Jaqueline had been assisting him, "a somewhat illiterate man," with his correspondence and accounts. Jaqueline replied that she would return if the Forestry Commission agreed she was needed. The next day a Commission official confirmed that her return was in the national interest, and he explained the situation to Rowse. [G.18]

Figure 4.1   Jaqueline with friends at Linwood in the early 1940s

"I am letting you down," Jaqueline apologized to Rowse. "I felt that ... my duty lay with the men whom I had helped to build up their little forest mill." Rowse tried to convince her—and the Forestry Commission—that she had made a mistake. He left the door open for Jaqueline to reconsider, and "to shoulder your share of the work of reconstruction." Meanwhile he hired Anthony Pott, of ACP, but told Jaqueline, Pott "is not fully trained, you are—nor has he that very necessary bent towards earthy things which you possess." [G.19, 20, 21] Jaqueline was relived to receive Rowse's letter. German air raids around Linwood had intensified. She promised to get in touch again if she could get "get someone who will conscientiously and lovingly take over this job," and volunteered to do whatever she could to help him in the meantime. [G.22]

Despite her devotion to the men at the mill, Jaqueline continued to be accused of spying, likely a symptom of the fear spread by the German bombs. When the local postmistress informed her that one of men at the mill was spreading rumors that she had regular rendezvous with a German soldier, Jaqueline angrily confronted the man. When he denied the lie, she "boxed his ear," dragged him down to the post office and got him to confess in front of two witnesses. [8/17/40]

Such indignities made Jaqueline more receptive to Rowse's appeals. "Clear up the immediate problem and then face up to the really big issue," he wrote on August 25th. "I will see if I can find another grant on which to bring you on ... to take full charge of the organization for South Wales." Rowse also stated that he was in contact with sociologist Karl Mannheim—then a lecturer at LSE—and hoped that would lead to some "real work" for Jaqueline. [G.23] PEP founder Julian Huxley (1940, 3) hailed Mannheim's recently published *Man and Society in an Age of Reconstruction* (1940) as "likely to become the planners' bible ... a compendium of modern sociological fact and theory ... woven round the central theme o the planned world's dawning." Jaqueline wrote Rowse: "I am most interested that you are in touch with Karl Mannheim," as she was then "working through" *Man and Society* as part of a program of study she had set for herself. [G.24] (This plan also included study of lichens and Quakerism.)

In late August, Jaqueline visited the Hammerschlags, who were having a hard time in Welwyn. "The 'Britons' won't be friendly with 'Aliens,'" she observed, and empathized upon learning that "the problem is mainly tiresome neighbours—local Col. O. business." [9/1/40] Her visit left her keenly aware of her own outsider status in Linwood. It also allowed Colonel Ogilvie to resume his attacks. "It appears that Col. O. has again taken advantage of my absence to raise a stink—& practically ordered [her landlord] to get me out of his cottage! ... It's an awful nuisance—I hate to 'make trouble.'" [9/3/40]

That same day she wrote Rowse offering to use her next "leave" to do some work for him in South Wales, and suggesting she would be available at the end of the year. A few days later she finally sent Rowse her Welwyn Survey report, hoping that it would "be submitted to the proper authorities as the Research Work that is still needed before I can receive the School's Pass Diploma."[G.25]

Meanwhile, the Blitz of London began in early September 1940, with heavy air raids targeting the working class neighborhoods of the East End. Jaqueline's thoughts were with the people of Stepney and her memories of her group of Girl Guides at Dame Colet House. While she had despised "that Polish Jewish crowd" in her youth, she now wished she could be there with them. [9/9/40]

Regional Planning for South Wales

James Forrester saw his collaboration with Rowse as a continuation of the work he had begun with Peter Scott (1890–1972) in Southern Wales. Scott was among a group of Quakers who went to live in the remote Welsh mining village of Brynmawr in 1926. They "believed in the need for radical change in the social and economic system and thought that this could be achieved through experiment" involving the local community (Baker 2004, 38). In 1934 they established the Subsistence Production Society (SPS) to help unemployed miners to produce goods for their own consumption and to own the means of production. Scott served as director of the Community House, which coordinated the exchange of volunteer labor for food grown on local farms. In 1936 Forrester, then age 25, became an SPS organizer for the Eastern Valleys and Brynmawr. With financial help from Lord Nuffield (William Morris), Forrester organized a number of workshops, a bakery, butchery, and dairy farm. As war approached, though, many of the unemployed miners found work in munitions factories, and at the end of 1939 SPS activities closed down.

Forrester and Scott then began discussions with Rowse about South Wales. Those conversations led to the establishment of the Association for Planning and Regional Reconstruction (APRR) as the research arm of the School of Planning, with Forrester as chairman and chief financier. APRR was intended to be the vehicle for the work that Rowse wanted Jaqueline to lead. Rowse wrote to Scott in August 1940 describing his concept of the Composite Mind—a team of specialists—as "the only way by which the master-craft of planning can efficiently be undertaken." He suggested that they "try to compose a "mind" to deal with South Wales," and that they begin by establishing "what is the diet which human beings, of each sex and various ages, should be given to make life worth living – not merely keep just sufficiently live to work." [G.26] Food shortages were among the British government's most pressing problems in the early war years.

As soon as Rowse heard that Jaqueline could be available by the winter, he sent her the "trial set up of a 'Composite Mind' for South Wales" for her comments. But he told her not to go there yet: "I want your brain, free of a tired body, making clear thinking possible." Moreover, it was too soon: "If Lord Forrester can energise Peter Scott sufficiently, the Board will come into being in a few weeks." In the meantime, he added several books on agricultural and international relations to her reading list.

To illustrate his developing idea of how a "Composite Mind" would work, Rowse likened the operations of executive and advisory planning committees and technical experts to mind/brain functions. "Perhaps in this haphazard way we may start this ball rolling." [G.27] Rowse tossed this ball to Jaqueline to carry forward.

Jaqueline fitted perfectly into Rowse's composite mind for South Wales. Reading her Welwyn Survey report, he was so impressed by her use of a priced optimum diet as a basis for defining a fair wage that he adapted it as the basis for the South Wales project. He commended the report as "first rate work of the greatest importance," and sent a copy to Forrester as a fund raising tool, but he made no mention of her Diploma. [G.28]

Hesitation

In mid-September, Jaqueline had to defend herself against yet another accusation of being pro-Nazi. The Chief Constable presented her with testimony from a local woman that the previous May she'd "reviled the King & Country - said Germany was bound to win & anyway we'd be better under a dictator!!!" The woman thought nothing of it at first, but Colonel Ogilvie convinced her to report the conversation to the police. "Luckily they didn't believe it – but there was only my word against hers." [9/16/40] Again, her ties to the local gentry likely smoothed the waters. At tea with Major Jarvis and his wife a few days later, she learned that Colonel Ogilvie had been reported to Jarvis as a possible spy! Whether that was true or said to console her, she found it amusing. [9/21/40]

In need of "a day or two in an affectionate atmosphere," Jaqueline spent a weekend with friends at Dartington, now the closest thing she had to a home base. When her bus back was cancelled—a common wartime occurrence—she hitched a truck ride and then walked several miles in the pouring rain to pick up her bike for a wet ride against the wind to arrive at the mill by 8:15 am, "tired & wet, but not very late." Life in the forest was making Jaqueline quite resilient. In mid-October she moved into "Broomy Cottage," a shack with a field and a horse barn. Despite primitive conditions—"no light, water, sanitation or heat. Earth closet, primus stove, oil lamps, wood fire," she reported: "I shall get along quite well I think." [10/13/40]

A few weeks later a long letter "about future social planning" arrived from Rowse. "His matter is always good, but his manner tiresome in the highest degree—However he touches on several of my pet subjects & I've rushed into a reply." [11[5/40] Rowse had heard from Scott, who had received comments on the outline proposal from several colleagues, who advised against trying to launch their scheme during the next six months. Instead, Scott and Forrester wanted Rowse to send a couple of researchers to Wales to "discuss the matter informally with interested people, and ... see if it would not be possible to study one or two specific problems and perhaps formulate a scheme more clearly." [G.29] Forrester pointedly asked: "Is the elusive Miss Tyrwhitt available or not?" Rowse re-directed Forrester's query to Jaqueline asking: "Is she?" His army medical inspection was pending, and if called up "you must take over," he wrote:

> I think you know my point of view well enough now, so it won't take very long
> for us to get things straight. The work you do can go towards your Diploma and
> Final qualifications ... so lets hear from you! ... Come on! [G.30]

Jaqueline was torn: "Whichever way I turn I feel conscience-stricken ...I know how frightfully important planning is – but how utterly impotent most of the work is likely to prove – just dashing oneself against concrete obstructions." [11/16/40] She learned that the situation at APRR was vague at that time, which reinforced her fears, and she was advised by economist David Owen, then general secretary of PEP (1940–1941), to wait for a position in the new "Ministry of Reconstruction" which Lord Reith was calling for—a suggestion she liked. Jaqueline wrote Rowse saying she had decided to stay put [12/4/40].

But after meeting with Rowse and Forrester in late December Jaqueline changed her mind again. Forrester made the difference. Jaqueline "quite candidly fell" for him. "I've said I'll go there as soon as I can get away." [12/28/40] Rowse and Forrester had written a statement of APRR's aims by then:

> The Association has been formed to make a practical contribution towards the
> solution of the many problems of post-war development and reconstruction. It
> is the belief of the Association that even a small group of people, thinking and
> planning constructively at the present time, may render practical service of real

value to the nation. The Association … endeavours in its work to keep itself free from the bias of party or creed, and as an association to make no criticism that is not constructive. The emphasis of its activities lies essentially on technical research into current and coming problems, and the results of its work are available to all. (APRR 1948c)

By then Lord Reith, head of the new Ministry of Works and Buildings, presented his recommendations for "Reconstruction of Town and Country." The 1940 Council welcomed Reith's proposals, but urged that plans for reconstruction be based on a survey of national resources and considered in relation to the distribution of population. To that end, the Council supported a Demographic Survey and Plan for Great Britain, conducted under the direction of Geography Professor C. B. Fawcettt, a follower of Geddes who, along with his staff, was evacuated from London to the University of Aberystwyth, which had become a center of the regional survey movement (Crone 1964). In January 1941, Jaqueline was offered a position on the project team, led by the émigré German architect E.A. Gutkind. Jaqueline liked the idea of working with Fawcett, and in Wales, but Forrester convinced her otherwise. It was understood, however, that she and Gutkind would coordinate their efforts. Equally important, the 1940 Council became interested in the work of APRR.

Part II
1941–1945

Chapter 5
Association for Planning
and Regional Reconstruction

APRR: Getting Started

Tyrwhitt arrived in London in February 1941 and assumed her position as director of APRR at a pivotal moment in British planning history. In January, the weekly magazine *Picture Post* had published its "Plan for Britain" sparking a national conversation about a better Britain after the war. *Picture Post* editor Tom Hopkinson (1970: 90) explained that the troops and the public needed to know what they were fighting and sacrificing for: "an idea of the country we wanted to make, and a passionate ... determination to make it." Immediately, Jaqueline was at the center of this conversation, as several of Jaqueline's PEP colleagues contributed to the Picture Post plan—a vision of a clean, healthy, modern community based on cooperation and close to nature.

Concurrently, Lord Reith had obtained Cabinet approval for his proposed approach to planning for post-war reconstruction. In February 1941, he announced that his work would proceed on the assumption that planning would be accepted as a national policy and that some central planning authority would be required. Reith assembled a small "Reconstruction Group" to conduct preliminary studies, which soon became the planning department of the Ministry of Works and Planning (Willatts 1987: 106–107).

At APRR's office in the Building Center, Tyrwhitt initially supervised a small group which included Anthony Pott, the eldest at age 37, Anne Radford (later Wheeler), 23, Peter Saxl, 26 and several members of the Architects Co-operative. Pott was about to leave for military service and welcomed her arrival and new ideas: "I still feel that given some organizing this show can do some useful work, but so far the lack of any direction, or even comprehensible idea, has been devastating in effect." [T.16]

Tyrwhitt didn't waste any time establishing contacts and lining up influential allies. She had a long talk with David Owen at PEP, which she saw as a model for APRR. She met with architect Judith Ledeboer, who was active in PEP and Honorary Secretary to the 1940 Council. Ledeboer had just joined the Ministry of Health, with responsibility for housing. Tyrwhitt renewed contacts at the Housing Center (which Ledeboer had co-founded in 1934); with her former AA classmate, Bobby Carter, now Royal Institute of British Architecture (RIBA) librarian; and with Pepler, now affiliated with Reith's Reconstruction Group [3-4/41].

In early March, Tyrwhitt sat down with Forrester and APRR board member architect Cyril Sjostrom to decide a research agenda. It included: regional planning; industry; agriculture and nutrition; services; population; housing and recreation; health and education; and uses of waste. APRR would follow PEP's example, and publish the results of its research in Broadsheet format, "designed as a suggested standard for the presentation of planning information" (APRR 1944). The idea of standardization to facilitate communication across specializations was a keynote of APRR's effort to create a "composite mind."

Tyrwhitt also began to coordinate with Gutkind's team in Aberystwyth; she would meet Gutkind for the first time at the end of March at a conference of the Town and Country Planning Association (TCPA)—the newly renamed Garden City and Town Planning Association. Osborn was running TCPA from his office in Welwyn Garden City, and using TCPA as a platform to make a bid for leadership in the reconstruction planning campaign (Hardy 1991, Buder 1990). Tyrwhitt cultivated her relationship with Osborn, with whom she often worked on weekends in Welwyn.

The TCPA Conference, which was on "Practical Problems of National Planning," presented a timely networking opportunity for Tyrwhitt, as it brought together nearly all the key actors in the burgeoning reconstruction movement. She scheduled a meeting the following week with two members of Reith's Reconstruction Group, architects William Holford (1907–1975), then Professor of Civic Design at Liverpool University, and John Dower (1900–1947), who was responsible for town planning in PEP, "to describe what we are up to." They both "seemed suitably impressed," she noted with pleasure in her diary. As if to confirm her arrival on the British town planning scene, that same day, April 3 1941, she finally received her TPI Associate Membership.

Furthermore, Tyrwhitt established a satisfying social life in London when she moved into an apartment with Alison Milne, a violinist friend from Dartington. Their apartment became the gathering place for an eclectic group, including Dartington friends, family, APRR staff, "Land Girls" from the New Forest, soldiers on leave and "random strays." She flourished in this atmosphere of camaraderie and in her new role as hostess.

APRR's office also became a hub of activity. Every Tuesday afternoon the Association hosted open discussions of work in progress. These proved to be immediately popular, and lasted into the night. Progress on research, however, depended on finding the right people interested in doing the work, often voluntarily, in their spare time. The staff now included Dr. Annie Noll, a Polish Jewish émigré. Tyrwhitt paired her to work with Saxl on health and education issues, which became early APRR focal points. APRR's work in those areas benefited enormously from collaboration with Dr. Innes Pearse (1889–1978), the cofounder, with her husband Dr. George Williamson (1884–1953) of the Peckham Health Centre. Tyrwhitt particularly enjoyed Dr. Pearse, who was "full of biological conceptions of planning." [4/21/41] The Peckham Health Centre was closed, as its modern glass-walled building was not safe during the blitz. APRR provided a forum for Pearse's ideas during the war, just as PEP had for Williamson in the 1930s.

Discussions at APRR were closely connected to those at PEP—where Pearse like Tyrwhitt was among the handful of active women members (Nicholson 1981: 39). Tyrwhitt was thrilled when her PEP colleague Oliver Roskill said he would also do work for APRR. [4/30/41]. The PEP Club provided a convenient place for Tyrwhitt to keep abreast of the latest developments in reconstruction planning, with regular presentations by group chairs, Members of Parliament and other personages (Lindsay 1981: 23). At one lunch talk in early May Reith reported on the work of his Reconstruction Group. In April two consultants to that group, geographers Dudley Stamp (1898–1966) and Professor Eva Taylor (1879–1966), had proposed that Reith initiate a National Atlas starting with maps important to planners. Reith authorized an advisory Maps Committee (including Taylor and Stamp) to list the maps needed. Dr. E.C. Willatts was assigned to take charge of this work within the Ministry (Willatts 1987, Campbell 1987).

In support of this initiative, APRR's first Broadsheet was to be on "The Delimitation of Regions for Planning Purposes"—although no one was yet available to work on it. Then, on May 10th, Taylor's maps were lost in an air raid that also destroyed the Building Center. Tyrwhitt quickly took matters in hand finding temporary space for APRR nearby, helping salvage APRR's things from the wreckage and depositing them in the new space, where work continued without interruption. [T.5]

Sadly, Saxl died during the air raid and Tyrwhitt picked up the work he had been doing for APRR on education. To get up to speed she spent a weekend with Dr. Pearse at her cottage in Somerset. They talked till midnight about the education report and about Tyrwhitt's new draft on Balance of Industry. The next day they talked while walking through the countryside. "All day was one long talk really – interspersed with appreciation of nature. I think I've got the education pattern … finally sorted out … . Also a relationship of health & sickness services & a method of their organization & general integration." The following day, as they hiked they discussed "a scheme of biological research, biological teaching & ethical teaching integrated–experimental stations outside towns as sites for weekend camps." Tyrwhitt returned to London energized and teamed up with Noll to develop those ideas. [5/16-8/42]

APRR now benefited from the support of Judith Ledeboer, who offered use of her former office in Russell Square. This move marked the beginning of "a period of great activity under wartime difficulties. Voluntary workers appeared from all sides," Tyrwhitt recalled. [T.5] She was also able to hire the Danish-born structural engineer Ove Arup to work on "District Heating possibilities," and a chemical engineer to work on "Sewage Disposal schemes." Tyrwhitt wrote in her diary on May 28th: "It's all very *interesting* but horribly *exhausting* …. However I think it's helping a movement towards a 'composite mind.'" [5/28/41] A month later, at Dorothy Elmhirst's invitation, Tyrwhitt, Forrester, Noll, and Arup spent a weekend at Dartington to think strategically about APRR's growth in consultation with the estate's experts. This retreat provided a timely stimulus to APPR at a crucial stage in its development.

Associations

Throughout the summer, Tyrwhitt enthusiastically built relationships between APRR and other like-minded organizations, notably the New Education Fellowship (NEF), a consortium of progressive educators, and the West Midlands Group on Postwar Reconstruction and Planning, a regional survey group based at University of Birmingham. Tyrwhitt conceived of a research project into "what makes a 'community'" with Eileen Thomas, a social welfare specialist affiliated with the Institute of Sociology/Le Play House—an organization formed in 1930 to promote Geddes's regional survey methods—in which she hoped to involve the West Midland Group, the Fabian Society, and the National Council of Social Service, an umbrella organization for voluntary groups. [8/11/41] These relationships flourished even if particular projects did not materialize.

Less enthusiastically, Tyrwhitt established ties to RIBA. She served on a RIBA "Planning Committee," but disparaged the other members as "old fools" who were "incredibly out of touch." Her rationale for staying involved even though she was doubly an outsider—a woman, and not an architect—was: "they *have* the prestige & it's *just* possible that one may be able to influence them enough to prevent the worst blunders." [10/3/41, 10/20/41] However, she did enjoy her friendship with Bobby Carter, spending hours talking with him at the RIBA library. Carter was very well connected and very willing to help both Tyrwhitt and APRR. He was an advocate for architectural research and worked closely with physicist J.D. Bernal (1901–1971) to establish RIBA's Architectural Science Board (ASB). Tyrwhitt became involved in the ASB's Education Committee through Carter, and, more broadly, in discussions of the role of science in planning.

Membership in the Women's Farm and Garden Association (WFGA) offered Tyrwhitt an opportunity to integrate her interests in agriculture and education, and acquire more experience in a leadership position. In July 1941 she was elected a member of the WFGA Executive Council, where she joined Brenda Colvin. They were part of a "new wave" of young, professional women who were revitalizing WFGA at a time of declining membership and financial problems (King 1999). As a Council member, Tyrwhitt helped finalize a scheme that Colvin had first proposed in 1940 to train women gardeners during wartime to specialize in food production. The six-month course began in the winter of 1942. This led Tyrwhitt and Colvin to collaborate on another educational project for landscape architects, who WFGA supported as role models. In the summer of 1941 Tyrwhitt, Colvin and others developed a provisional syllabus, which notably included a component on ecology, as part of the Institute of Landscape Architects' (ILA) effort to produce qualified practitioners and to establish the profession (Gibson 2001).

Maps: Ground Plan of Britain

After Germany's attack on Russia in late June 1941 diminished the danger of an invasion, the land market began to show signs of recovery, and people concerned with planning began to discuss the question of post war rural land use. Those discussions led to appointment in August 1941 of the Scott Committee on land utilization in rural areas (Ward 2004: 82; Sheail 1997: 389). Prior to that the 1940 Council's Map Committee resolved to produce its own set of planning maps to inform that debate. Initially, the committee paid for one mapper, who worked in the APRR office, beginning in mid-July 1941. A week later the Committee asked Tyrwhitt to take over the "Map Bibliography." [7/21/41] This was APRR's first contract assignment.

Fortuitously, that summer, Prof. Taylor agreed to join the APRR board. Tyrwhitt took an early morning train one Saturday and hiked over some hills to work with Taylor, who had evacuated from London to her cottage in the Cotswalds. Tyrwhitt was nonplussed to make such an effort as she considered the older woman "superb in every way." After a day-long discussion of APRR's maps Tyrwhitt wrote in her diary: "Think I've now got the whole series worked out—& plenty of other ideas beginning to ferment in corners of my mind." [8/2/41] Taylor now joined the list of unusual women Tyrwhitt sought out as mentors, including Innes Pearse, Ellen Willmott, and Chrystabel Procter.

The first set of maps (all drawn to a uniform scale of 10 miles to the inch) had to be done by the end of September. Tyrwhitt immediately offered a draftsman job to the refugee German architect Wolfgang Frankl, who had been referred by Gutkind. Tyrwhitt liked Frankl "at first sight." [7/19/41] He immediately liked her, too. "Then she introduced me to all of them, her family," Frankl recalled. "During the war she collected stranded people and put them to use." [E.2] Tyrwhitt wanted to hire some of Gutkind's staff as the Demographic Survey group's work was ending in September. APRR could then "take on the National Atlas scheme that's been hanging fire so long." [8/25/41] On her way back to London after visiting her sister's growing family in Wales, she made a detour, involving a seven-hour train ride to Aberystwyth to meet with Gutkind's team. She concluded they were "not up to much." [9/1/41] But her dismissive comment says more about the rivalry that existed between her and Gutkind than his work. This rivalry may have been due in part to a clash of personalities, and in part to Tyrwhitt's insecurities, which fueled her tendency to be critical. She admitted: "I know he looks at things the right way—but I *couldn't* work with him. He is too inflexible—too sure." [10/23/41]

When Ledeboer, Pepler, Stamp, and Jellicoe met at APRR's office on September 4th to review the work in progress on the maps, Jaqueline crowed that they were "suitably thrilled – & so they should be!" The committee decided to pay APRR until Christmas to prepare another set of maps. In this way APRR was in the forefront of a broad movement to provide a scientific basis for planning by presenting facts in a visually unbiased way. The London *Times* supported this effort in an editorial: "Much lip service is paid to the idea of planning, but it is not always realized that planning will be neither effective nor tolerable unless it is backed by science ... . Science in its turn must increasingly become the servant not of war, or of big business, or of a particular regime, but of the general welfare of mankind" (Science 1941: 5). This editorial was prompted by the international conference on Science and World Order, hosted by the British Association in late September 1941, which brought together scientists, civil servants, and elected officials. Two APRR members, Arup and Noll, were among the speakers. "Pretty good for so young an organization as ours," Tyrwhitt noted. [9/27/41] In his remarks Arup articulated APRR's approach:

> The problem is the same here as in other spheres of human activity—a wealth of new knowledge, new materials, new processes has so widened the field of possibilities that it cannot be adequately surveyed by a single mind ... . This produces the specialist or expert, and the usual problem arises, how to create the organization, the "*composite mind*" so the speak, which can achieve a well-balanced synthesis from the wealth of available detail. [emphasis added] [in Gutkind 1944, 58]

A few weeks later, the 1940 Map Committee met at APRR to review the first set of planning maps and to discuss publication and the schedule for producing the next group. This publication was to be called a *Ground Plan of Britain*. Tyrwhitt commented that the committee "fell all over the maps—except Eva Taylor! That grand woman decided to put me in my place & I got properly ticked off about one or two things – none of them my fault but all great fun." While Taylor's often tactless manner alienated some, her criticism inspired Tyrwhitt to work harder. In a way, Tyrwhitt, then 36, and Taylor, then 62, were kindred spirits. "Jacky was a forceful personality, full of ideals and full of energy, straightforward and without fear," observed Frankl, who found his occasional disagreements with her stimulating. "I liked the way she came along and told me what to do, I felt privileged – without reason; she treated everybody the same way." [E.2]

Evidence for the Scott Committee: Sustainable Food Systems

For the remainder of 1941, the work of APRR focused on producing the *Ground Plan of Britain* and preparing evidence for the Scott Committee. Tyrwhitt led the process of developing APRR's evidence report, which was based on a series of broadsheets that she and Forrester prepared on "the consumption, production and distribution of those perishable foodstuffs (other than butter and fish) which can easily be produced at home" (APRR 1941a). Tyrwhitt was the principal author of Broadsheet No. 4: "Consumption of Fresh Food," which built on her earlier work for the Welwyn Survey on a priced optimum diet. This Broadsheet called for Britain to adopt a balanced diet and to promote local self-reliance in food production by local "service" agriculture, located close to the point of consumption. Such a service agricultural industry would require about 54 percent of the agricultural land of the country, allowing the remaining 46 percent to be available for a more distant "supply" agriculture. Forrester drafted Broadsheets No. 5: "Production of Fresh Food," which developed the case for a service agriculture; and No. 6: "Distribution of Fresh Food," which outlined a system of fresh food production and distribution assuming a network of "Control Farms and "Fresh Food Centers." The centers would function as cooperatives, and "care for the welfare of their staff," thus modeling a "third way" between government ownership and free enterprise (APRR 1941b, 1942a).

As part of their investigation of scientific approaches to farming, Forrester arranged for Tyrwhitt and Roskill to spend a weekend with him in South Wales. They spent Saturday with Peter Scott at one of the Brynmawr Subsistence Scheme farms. On Sunday they visited ex-miners and their wives. Then Tyrwhitt and Forrester met with Lady Eve Balfour (1899–1990) to discuss her experimental organic farm near Ipswich, and visited Sir George Stapledon (1882–1960)—Director of the Welsh Plant Breeding station (1919–1942)—at his farm near Stratford on Avon, where he was testing crop rotation

To help produce APRR's evidence report, Tyrwhitt hired émigré architect Gerhard Rosenberg. After pulling together Rosenberg's work and her own she went to work with Forrester at her apartment to hammer out a draft over the weekend. "We both typed hard till lunch, discussed our findings over coffee, worked again at out typewriters till about 3.30. Then joined forces and started editing. We got through a good deal—both of us being as critical as we knew how … . He is grand to work with." [11/9/41] Indeed, the more time Tyrwhitt spent with Forrester, the more she admired him. She confessed: "I've never yet felt so completely at ease with anyone mentally," but "have not yet been able to call him 'Jim' to his face." [7/28/41] Aside from being his subordinate on the job, she was very aware that he was a peer and she was not.

Tyrwhitt and Forrester presented their draft for feedback the following Tuesday at APRR's regular discussion meeting. "It's been a day I shan't forget,"

> Everyone in London who is likely to cut any ice in the planning world was there except the 3 actual officials who (naturally) couldn't be; but we know they are for us! It was a grand show – just grouped around the room on stools & really getting down to the job of vetting the report. Then – in the pouring rain – several of us fed at the Lyons [café] & after that Jim & I got down to work. I induced him to come to the flat as I was wet through. We worked till 2.30 am & he is now sleeping in my flea [sleeping] bag. [11/11/41]

Rosenberg, who was part of many such exhausting after-hour work sessions attested: "Jaqueline's obvious willingness to work as hard as anyone whom she employed, and the fact that she had no axe to grind and no wish to gain any advantage for herself, gained her the respect and in the long run the affection of the many people of all backgrounds and positions with which she came in contact." [E.3]

Tyrwhitt then worked night and day preparing the text and graphics to include in the final evidence report. (It is interesting to note that Broadsheet No. 4 "Consumption of Fresh Foods" is an early example of APRR's use of pictorial statistics, which were popularized by social scientist Otto Neurath—an Austrian émigré who had recently settled in Oxford—to make social facts easier to understand.) The Scott Committee devoted all day December 5 to hearing APRR's oral evidence. The APRR team included Forrester, Tyrwhitt, Professor Taylor, Dr. Noll, and Roskill. "Jim & I bore the brunt & Oliver & Prof. Taylor backed us grandly. Annie hadn't much to say—but she said it well," Jaqueline reported. "Prof. Dennison [economics] was our bitterest enemy—& Dudley Stamp was antagonistic to the farming policies. But it was a good show on the whole." [12/5/41]

FOOD REQUIREMENTS PER 1,000 PEOPLE PER DAY

The foods shown below could be produced at home in zones of service agriculture, but include only the minimum weight of meat [see diagrams page 3 and in Broadsheets 5 and 6.]

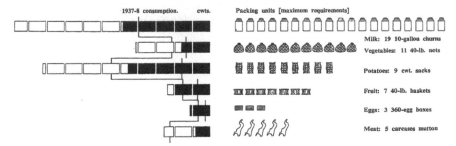

Most of the foods shown below would have to be imported, including the rest of the meat.

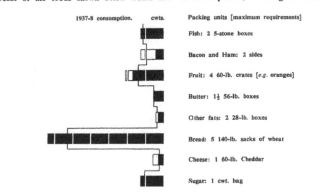

Each rectangle equals 1 cwt. Black, minimum requirements; White, maximum requirements. When there is a second line, in addition to the line indicating 1937 consumption, this shows the proportion imported out of the total consumed in that year. These imports have only been indicated in the first group of foodstuffs.

Packing units named are representative, but vary locally and with means of transport.

The diagram shows that the fresh foods and bread present the major handling problems, by reason of their weight and bulk.

The Association for Planning and Regional Reconstruction welcomes comment or criticism on this or any other aspect of its work. No charge is made for Broadsheets, which are sent free on request to all interested, but the Association will be grateful for contributions towards the cost of their production, which is considerable.

Figure 5.1   A page from APRR Broadsheet No. 4 "Consumption of Fresh Food" (1941) exemplifies the Association's effort to visualize information, an abiding concern

Recognition

One sign of the respect Tyrwhitt had garnered after only nine months at APRR was an invitation from economist G. D. H. Cole (1889–1959) to attend a private conference at Nuffield College in Oxford in December 1941 on Practical Aspects of Post-war Building; Forrester also attended. The meeting took place just after the entry of the US into the war lifted the mood in Britain. It was the second in a series (1941–1943) that Cole chaired, intending to shape a consensus around the aims of post-war reconstruction, and a collectivist "middle-way" to achieve them (Ritschel 1997: 316). Tyrwhitt was "staggered" to be one of only three women guests; the others were Prof. Taylor and Hermione Hitchens (a member of the Barlow Commission and the Scott Committee).

That weekend in Oxford was also a memorable one for Tyrwhitt because she got to see the *Ground Plan of Britain* page proofs; she, Forrester, and Taylor joined Jellicoe in correcting them. The 14 maps included in this book partially realized Taylor's desire for a National Atlas as a basis for post-war construction (APRR 1944).

After the Christmas holiday, which Tyrwhitt spent with her father (the first time he had ever invited her to a meal), she visited Taylor, her new mentor, in her "wee primitive cottage" in the Cotswalds, and went with her to Gloucester to see architect Max Lock, who was then Head of the Hull School of Architecture, and look over survey maps with the County Planning Officer. Tyrwhitt had earlier discussed a partnership between APPR and Lock based on a proposal he had made "to set up and launch a contemporary version of Geddes's ... neglected methodology of Civic Diagnosis and Civic Surgery" in Hull, which was a heavily war damaged and badly blighted city region. [E.4] That partnership never materialized, but inspired by Geddes's ideas, Tyrwhitt and Lock began a collaboration to make surveys more useful analytical tools for physical planning.

Catalyst and Synthesizer

Lock had been a student at AA a few years after Tyrwhitt. When they first met, sometime later, he had asked her "why she had never set up in practice," Lock recalled. "She replied that she had come to think of herself as a 'catalyst' rather than as a practicing architect or town planner, one who, while not herself changing, makes vital chemical changes in others. 'You mean,' [he] asked, 'like Geddes?'" [E.4] In less than a year Tyrwhitt had, indeed, served as the catalyst that crystallized APRR into the "composite mind" that it had become. But in addition to organizing and inspiring the work of others, Tyrwhitt's own work researching and writing broadsheets, collecting, analyzing and visualizing statistics, and conceptualizing maps led her to begin to develop her own ideas about regions and decentralization. She had rehearsed those ideas in a series of lectures she gave to the Workers Educational Association (WEA) in Ipswich in September and October 1941. The success of those lectures encouraged her to have confidence in her own contribution to the synthesis of ideas she was forging. "I must say I like my audience," she wrote. "It's fun talking to them & I know what I say is good—50% Rowse, 30% Forrester & 20% me." [10/6/41]

In late December 1941, Tyrwhitt dined with her old friend Jack Blaksley who was now assigned to the War Office Directorate of Education. "Jack is off on his hobby horse. Aristocracy to the fore," she observed disparagingly. "He's an excellent Nazi at heart & while patriotic to his toenail is wrong—dead wrong—in his philosophy. It's the short term easy way out—the *Führer princip* & its bad to have him on Army Education." [12/26/41] If Blaksley and Roskill had represented the two poles in a spectrum of ideas that attracted Tyrwhitt in 1931, she was now clearly identified with Roskill. She had completed her personal and intellectual transformations, spurred by the profound impacts of the war.

But on New Year's Eve, Jaqueline was alone. "It's strange how difficult I find it to grasp that I am just undesirable except as a working companion—and I've never really been able to discover why," she admitted to her diary. "I know I don't take much trouble about my appearance but it's not too bad most of the time—but of course I am horribly bossy, horribly critical & I think I never admire people enough—superficially."

She paid a price for traveling and working non-stop—and for being so hard on herself (and others). After one particularly grueling trip she was so worn out that she had to stay in bed for days. "Oh damn damn damn my constitution. Though I know it's 'all my own fault,'" she confessed. "That is I know I've abused it—but it's had to be abused—overworked—in order to compensate for a natural life. It's the only thing that has kept me unbitter." [9/8/41] But Jaqueline did not indulge in self-pity for long. She was too full of energy, and she found the work "thrilling" even as it "piled up 10 deep." [9/3/41] It's a testament to her creative energy that she was able to rise above self-defeating thoughts and dedicate herself to the cause of post war reconstruction.

Chapter 6
Composite Mind:
Planning for Post-war Reconstruction

Group Work

APRR was founded on the principle of collaboration, but since money was tight an enormous share of the Association's work fell on Tyrwhitt's shoulders. Annie Noll was sufficiently concerned about the situation to confront Lord Forrester and urge him to take on a more active role in APRR affairs in order to relieve Tyrwhitt of some of the burden. Noll knew that episodes of heart trouble Tyrwhitt had recently endured bespoke the strain she was under. But Forrester's father, the Earl of Verulam, objected to his son spending any more time in this side venture. As a result, Forrester continued to limit his involvement at APRR to the weekly discussion meetings, often arriving late, and the smaller group sessions that continued at Lyons Corner House after the office closed. Too often this meant that Tyrwhitt had only a few minutes alone with Forrester, on the platform before his last train at midnight, when they would go over the business of the week. This left her feeling "tired beyond measure &—probably consequently—depressed." [2/5/42]

On the other hand, Tyrwhitt probably found the hectic pace of work a welcome distraction from worrying about her brother Cuthbert and sister-in-law Delia in Singapore, which, in January 1942, was threatened by an imminent Japanese Army invasion. It seemed to Tyrwhitt that among her family, Cuthbert was "the only one left who mattered" [1/31/42] In mid-February, in anticipation of the fall of Singapore, Forrester counseled Tyrwhitt: "There is nothing to do but hope & work" [2/13/42]. He urged her to take April off and spend time in Scotland at his expense. But Tyrwhitt felt her place was at APRR.

Aside from work, Tyrwhitt found comfort in choral singing, having re-joined the Bach Choir after a seven-year absence. The experience of performing with this choir exemplified her ideal of collective effort: "140 people—all voluntary & all knowing their jobs." [1/4/42] Likewise, more intimate music parties at her flat, with friends from the office and Dartington, were restorative and a model of harmony achieved through cooperation. Tyrwhitt loved working with Forrester because she felt they had achieved a similar accord. "It's so grand to see your own ideas emerging in someone else's words—suddenly grown & clothed and mature, when you'd left them wee and naked." [1/27/42]

Surveys and Maps for the National Plan

Tyrwhitt was eager to get on with the new work at APRR: producing a new series of maps, mostly on aspects of "social health." [1/6/42] Her thinking about survey methods and mapping techniques was stimulated by Pepler, who had pioneered the application of Geddes's principles of regional survey in the field of planning, just as Taylor had in the field of applied geography. To that end, Pepler (1923: 70) had long advocated a standardized system of notation for civic surveys and town plans along the lines APRR was experimenting with. Likewise H. V. Lanchester (1863–1953), Tyrwhitt's colleague on a RIBA Reconstruction Subcommittee)—who had worked with Geddes in India—called for a standardized notation system for graphically displaying survey data on thematic maps. Lanchester's (1915) efforts during World War I to engage unemployed architects to conduct "civic development surveys" to prepare for post-war reconstruction can be seen as a forerunner of APRR's work.

Tyrwhitt carried this cartographic tradition forward by connecting APRR's work developing cross-disciplinary survey techniques to the broader discourse on scientific planning. She felt honored to be invited to a small conference—where she was the only woman—organized in January 1942 by the British Association "to try and inject a scientific ethos into social and economic planning" (Wersky 1978: 271). Pepler was enthusiastic about APRR's new maps—another step towards realization of Taylor's National Atlas—and spoke about them at PEP in March 1942. Tyrwhitt partnered with RIBA Librarian Bobby Carter to organize a meeting there to systematize the scope and content of the burgeoning wartime regional survey initiatives. Alexander Farquharsen (1882–1954), director of the Institute of Sociology chaired the meeting, which took place in June 1942 and was attended by nearly 40 people from different survey organizations.

In early 1942 APRR's small team of mapmakers included two artists Tyrwhitt had met through the Land Army Forestry Service, Kitty Boole and Bunty Wills. Their artistry was enhanced by the contribution of the émigré German mathematician Matthias Landau, whom Tyrwhitt hired in May 1942 to work with APRR and PEP. [5/4/42] Thanks largely to Landau's input, by September 1942 Tyrwhitt felt confident that APRR's new series of maps were good enough to publish. [9/9/42] In conceiving this publication Tyrwhitt came up with the idea of relating the maps to the Barlow, Scott, and Beveridge reports—published in 1940, August 1942, and December 1942, respectively—the key documents which provided the framework for the British government to undertake comprehensive planning. On December 8th Tyrwhitt happily reported in her diary that this idea was "very well received" at a meeting at APRR where the new maps were displayed and "several government departments were represented." With this validation she confidently approached Lund Humphries, who quickly agreed to publish it.

Maps for the National Plan: A background to The Barlow report, The Scott Report, The Beveridge Report (1945a), credited collectively to APPR, featured 36 different maps of England, Scotland and Wales, all but two drawn to a common scale. Two transparent overlays were provided showing chief urban areas and county boundaries, and the distribution of population. The maps employed a black and white notation system for ease of reproduction that was notable as an ambitious attempt at the visualization of the spatial distribution of complex sets of socioeconomic data.

Tyrwhitt enjoyed inventing new maps; she listed "map making" as a hobby on a resume she prepared in 1945 [T.17]. One challenge Tyrwhitt and APRR faced in making maps to support national planning was how to delimit the boundaries of

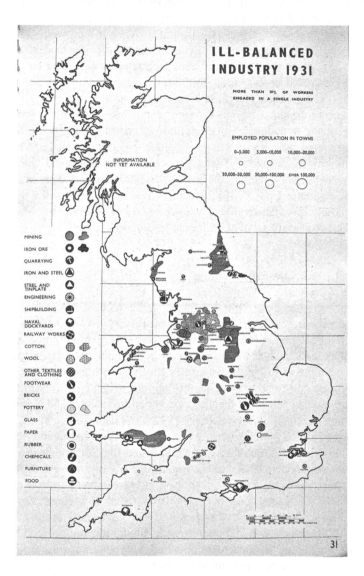

Figure 6.1   An example of APRR's innovative technique of mapping complex social statistics

regions to serve as planning units. APRR used the concept of a functional region based on the "principle of wholeness, of integration of people, place and work"—along with "the cell principle of aggregation of small units into regions of different orders"— that had been developed by, among others, the French geographer Le Play, Geddes, and Christaller. Tyrwhitt and her colleagues thought that regional planning authorities bounded in those ways could feasibly mediate top down and bottom up planning in a small, insular country such as the United Kingdom (APRR 1942c: 4, 1942d: 1).

A good example of Tyrwhitt's inventive map making is the set of maps she devised for the National Federation of Women's Institutes. Tyrwhitt conceived of these maps, which were based on a survey of local cooking customs in England and Wales, as part of a series "which show many different aspects of the life of the country which need to be considered in 'planning'" (APRR 1942e). Tyrwhitt's idea was that the incidence of cookery customs might indicate areas of local homogeneity, and suggest the boundaries of a *cultural* region. Tyrwhitt reported that a survey of recipes made at home according to traditional methods showed "that the incidence of local food customs were more closely related to geographical boundaries than to the boundaries of the administrative counties." The maps demonstrated that some "local" dishes had become almost universal; in other words, traditional recipes were not a reliable indicator of cultural regions. [T.18] Tyrwhitt's food maps impressed the American cultural anthropologist Margaret Mead, who visited APRR in October 1943. At Mead's request APRR produced a set of maps for her to take back to the US. [10/23/43]

Food, Health and National Planning

The war galvanized many people into taking action on the issue of the nutritional state of the nation. Scottish scientist Sir John Boyd Orr (1880–1971), who had emphasized the relation of poverty and poor food to ill health in Food, health and income (1936), took the lead in organizing the Nutrition Society in 1941 to provide a forum for people of different disciplines involved in nutrition work "to share information and constructive criticism" (Copping 1978: 110). Tyrwhitt attended the Nutrition Society's second meeting, on "Food Production and distribution in relation to nutritional needs," in February 1942. Orr's paper there, on "Agricultural implications of a food policy based on nutritional needs," bolstered APRR's case for a service-oriented agriculture. Tyrwhitt found the Society's third meeting, on "Problems of collective feeding in war time," equally stimulating. [5/30/42] Lord Woolton, Minister of Food, described the government's plans to provide communal meals at canteens in factories, restaurants, and schools "to keep the people healthy and ready for the work required of them in wartime and afterwards" (Copping 1978: 118). By then Tyrwhitt was already planning to conduct a full scale experiment along the lines recommended by Woolton, thanks to Forrester.

Tyrwhitt told Forrester early in May 1942, that she wanted to do something practical "as a sideline." He assigned her the job of getting a market garden going to provide food for the Enfield Cable Works canteen, which served up to 1,500 meals a day. [5/7/42] Tyrwhitt spent all available free time through June 1942 devising the cropping scheme and organizing a staff of nine Land Army workers (Land Girls) to cultivate the twenty-two-acre garden near the Enfield factory in St. Albans (Enfield 1943). Five additional acres were cultivated for fruit in the fall. Tyrwhitt benefited from the advice of her old friend, Chrystabel Procter, who, as garden steward at Girton College, Cambridge, during the war, supplied the college's kitchens and Cambridge school canteens with much of their fruit and vegetables (Bailes 2004: 455).

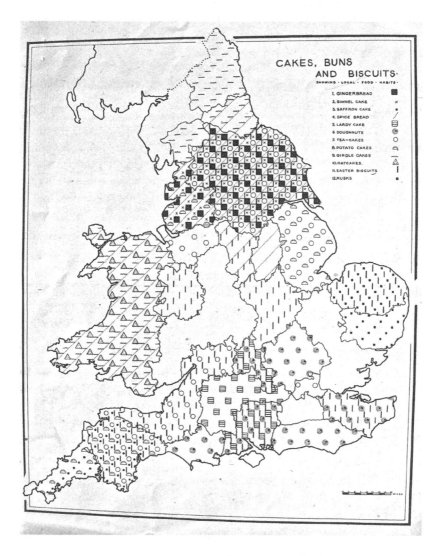

Figure 6.2   Tyrwhitt devised this map of cookery customs to investigate cultural regions

Tyrwhitt volunteered along similar lines as a member of the WFGA Council (which she supported for sentimental reasons: "loyalty to the kindness & charm of Ellen Willmott—one of the few women I've loved") [4/15/42]. Among other things Tyrwhitt organized a WFGA conference in November 1942, at Old Lye House, a facility owned by Forrester's family, to train young women gardeners to specialize in food production. At a concurrent WFGA conference for regional organizers in London Tyrwhitt joined Lady Eve Balfour, who returned with her to Old Lye House to speak to the group on soil fertility. "Grand stuff that made the girls suddenly see life whole," Tyrwhitt thought. [11/7/42] At that time she and Forrester were working with Balfour on a study of farm management and suggestions for using human wastes to maintain soil fertility.[1]

Food preparation was also the focus of "The Hub of the House," a two-part report dealing with planning for kitchens from the user's perspective; APRR submitted this report as evidence to the Dudley Committee on the Design of Dwellings, which was established by the Ministry of Health in March 1942, with Judith Ledeboer as secretary. Part I, "The Town Kitchen," presented "designs suitable for low cost mass production" (APRR 1942b: 1) It was produced with financial and technical assistance from the English Joinery Manufacturer's Association and Tyrwhitt met with several architects and builders in order to convince them to incorporate its standardized designs as components in the large-scale housing projects for war workers then being built in Coventry.

Part II, "The Country Kitchen," addressed primarily the segment of the rural population who worked in rural industries, and surveyed all factors to consider in housing rural workers (APRR 1942f: 1). The varying geography of regions supporting rural industries made standardization of worker housing problematic. "The Country Kitchen" noted that in contrast to the town, "the amount of work done in the [country] kitchen is greater, and all the other domestic activities – washing, laundry-work, cleaning, heating and lighting – are in some degree connected to it." "The Country Kitchen" clearly reflected Tyrwhitt's concern with the needs of women who worked on the land.

Tyrwhitt was working concurrently with Dr. Pearse on her report for APRR: "Health and the Future."[2] Fresh, nutritious food was an important component of Drs. Pearse and Williamson's approach to promoting positive health, as opposed to simply curing or preventing disease. Dr. Pearse presented the results of their work, theorized as a biological study of the human organism in its environment, at

---

1   APRR, "Human Wastes, Their Collection and Disposal," Broadsheet No. 7, and "Human Wastes, A Summary of Existing Methods of Disposal," Broadsheet No. 8, were published before April 1943. They were renamed "Habitation Wastes," Part I and II; both are rare if existant.

2   APRR's procedure was for people working on topics to present a preliminary paper at one of the weekly discussion meetings. Based on this discussion the paper would be re-drafted as a report and circulated to a wider group for criticism. Reports that represented the views of APRR were condensed and published as four-page broadsheets (APRR 1943: 2–3).

APRR in December 1943, shortly after the publication of her book *The Peckham Experiment: A Study in the Living Structure of Society* (1943). Tyrwhitt revised Pearse's APRR report, and Pearse then rewrote portions. Tyrwhitt took note of one of Forrester's comments on the work in progress: "He picked on a para [sic] about 'we can only change social conditions so that the process of natural selection can operate more accurately' & pointed out that natural selection was not based on purpose." [11/27/42] His skepticism focused Tyrwhitt's attention on deriving practical lessons from their work.

Transition: Personal and Professional

At Easter 1942 Tyrwhitt was feeling uprooted, having just received notice that she and Alison Milne had to move. They soon found a larger, but shabbier apartment in North London. They fixed up the place themselves; Tyrwhitt was particularly proud of how they planned their own kitchen—a practical application of her work on the Hub of the House. She moved additional furniture and dishes from her cottage at Welwyn. Surrounded by more of her things, the new place soon felt like home, and was often filled with friends sharing meals, making music, or finding a bed for the night. Tyrwhitt thankfully affirmed: "I have the reality of the job, a home with Alison & a new friendship with Bunty [Wills] that keep me sane." Her relationship with Bunty, which bridged their professional and private lives, was especially nurturing. In many ways Alison, Bunty, and Annie Noll formed the core of a new surrogate family for Tyrwhitt, although unlike the early 1930s, this time the members were more like sisters than father or grandmotherly figures. As noted earlier, Tyrwhitt also treated her staff as family, taking care of them, especially the refugees, in a variety of ways. The feeling was mutual; her staff also took care of her after crushingly long days on the job or on those rare occasions when she was so ill or exhausted that she was forced to stay in bed.

Tyrwhitt was comfortable with her independence, but still longed for the intimacy of parenting and the love of a life partner. She lavished affection on her nieces and nephews in Wales, and her god-daughter, and helped pay for their education. Tyrwhitt found nurturing others a substitute for the love that eluded her. She admitted that "there's always been someone for whom I'd willingly die." [11/16/42] Currently it was Forrester, a remarkable yet repressed man whom she wanted to help. Tyrwhitt struggled to prevent her romantic feelings for him from interfering with their professional relationship.

At the same time that Tyrwhitt had to find a new place to live, she was also was looking for a new office space for APRR, as Ledeboer's lease was expiring. "By this time the Association had several contract jobs that had enabled us to build up a small paid staff, and with the backing of Jim Forrester we took the ground floor of No. 32 Gordon Square," she recalled. [T.3] But Tyrwhitt took it upon herself to look for funding so Forrester would not have to keep footing the bill. She first explored an alliance with the National Council for Social Service (NCSS) with whom APRR was sharing office space. Such a partnership was a good fit, as both APRR and NCSS emphasized the value of citizen engagement through voluntary cooperation rather than dependence on state action. Moreover, they both employed survey techniques towards complementary objectives, physical planning for APRR, and social welfare for NCSS.

NCSS staff introduced Tyrwhitt to one of their funders, a developer of industrial parks. She hoped he would back the new school of planning that APRR's board had asked her to organize. Instead he wanted APRR to do a job: design a prototype for converting military sites to post-war uses. This contract enabled Tyrwhitt to keep APRR's mappers employed, and to hire geographer AE Smailes, from Gutkind's staff. Under Tyrwhitt's supervision, the APRR team approached this survey from a regional planning perspective that integrated the new industrial site as a part of a comprehensive plan for the growth of an adjacent town, accommodating housing, open space, and agriculture.

The Neighborhood Unit

The contacts Tyrwhitt established with NCSS stimulated her thinking about regional planning based on the cell principle at a new scale: the "neighborhood unit," a residential area centered on civic facilities. NCSS was an important source of information in Britain on the neighborhood unit concept, which, while well-established on the Continent and in the US, was only then coming into vogue in Britain (Clapson 1998). Many modernist social housing projects incorporated the neighborhood unit to facilitate communal life. Refugee German architects, who settled in England in the 1930s and joined the MARS group, subsequently developed the neighborhood unit in the British context, notably in the MARS plan for post-war London, which was published in *Architectural Review* in June 1942 (Bullock 2002, Gold 1997). That month Tyrwhitt met with the principal authors, architect Arthur Korn (1891–1978) and engineer Felix Samuely (1902–1959) to explore collaboration between APRR and MARS. An affinity between CIAM urbanism and APRR's Geddessian approach provided a basis for teamwork (Korn et al 1971). Tyrwhitt's experience in Germany helped her establish a rapport with those men, particularly Samuely. [9/3/42] She began attending meetings of the MARS Town Planning committee and they participated in APRR's discussions. This led to Tyrwhitt's election as a member of the MARS group in early 1945. [T.19]

Tyrwhitt also developed a neighborhood unit project with the German refugee sociologist Ruth Glass (1912–1990). Sir Wyndham Deeds, vice chairman of NCSS advised on the project, a survey of the Bethnal Green section of London, where he lived. HG Stead, a leader of New Education Fellowship, who now worked with APPR, consulted on the integration of education in neighborhoods. [10/13/42] In orchestrating this survey, conducted between 1943–1945, Tyrwhitt led APRR into systematic investigation of the social basis of town planning.

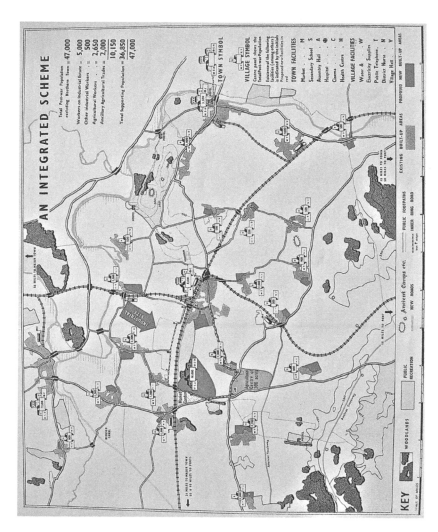

Figure 6.3   APRR proposal for reuse of a wartime site integrating agriculture and industry

Chapter 7
Training Planners for Post-war Reconstruction

A New School of Planning

Even though she was already overworked, in December 1941, Tyrwhitt reluctantly accepted the APRR board's request that she organize a correspondence course in town planning to carry on the work of SPRND during the war. When she presented APRR's proposal to the TPI Heads of Schools Committee in late January 1942, she proudly noted that she was not only the sole woman there, but also "the youngest by about twenty years." The TPI committee accepted the idea in principle, "coupled with a short studio and survey course after the war before [students] could qualify" for TPI membership. "One can't teach survey work by post," Tyrwhitt agreed. [1/30/42] There were still many substantive, legal and financial details to work out. Luckily Tyrwhitt had a powerful ally in Pepler, who then chaired the TPI Joint Examination Board.

Tyrwhitt worked with Pepler to re-design Rowse's prewar school to run both the wartime correspondence course and the post-war completion course. Progress setting up the new School of Planning and Research for Regional Development (SPRRD) was slow. Architect William Holford (1907–1975), who worked with Pepler at the Ministry of Works and Planning, agreed to serve on the school's board of advisers, as did W.A. Robson (1895–1980), a leading authority on metropolitan government. Tyrwhitt's old AA friend, W.F.B. Lovett, accepted her invitation to become Director of Studies—a job she did not then want. Tyrwhitt found the counsel of Dr. H.G. Stead to be "extremely helpful & illuminating … . My mind gets cleared & stimulated on lots of things when he is around." [12/1/42] Stead's recently published book, *The Education of a Community* (1942), analyzed the defects of Britain's educational system and outlined his vision of reform; his input located discussions about APRR's new school in the broader context of the educational reform movement in Britain at that time.

Dr. Stead chaired the first meeting of the SPRRD advisory board in August 1942. The institutional connection between SPRRD and APRR was clear: two out of six school board members—Prof. Taylor and lawyer Ambrose Appelbe —also served on APRR's board; and Tyrwhitt played a leadership role in both. Tyrwhitt was charged with drafting a summary of the school curriculum and the Association's work and survey procedure to provide a basis for discussion with potential funders. [8/28/42]

Army Education

The timing was ripe for APRR to launch the correspondence course. By then the Army had concluded that the voluntary educational program for enlisted personnel that started in February 1941 was insufficient. A new branch of Army Education was created in August 1941, the Army Bureau of Current Affairs (ABCA), marking the introduction of compulsory education (Hawkins and Brimble 1947). Echoing the idealism of the *Picture Post's* "Plan for Britain," ABCA aimed to give the troops an opportunity to discuss current affairs to foster the understanding that they were fighting for something beyond national survival—the building of a better Britain. The ABCA program was well established by early 1942, publishing a weekly pamphlet *Current Affairs* framing topics for discussion. A correspondence course in Town Planning would appeal not only to men and women who intended to become town planners, but also to anyone who wanted to take part, as a responsible and informed citizen, in the physical reconstruction of Britain.

By the summer of 1942, planning for peace by the government was gaining momentum and military progress allowed the extension of correspondence courses to more troops. The Army Education Corps also increased the scope of compulsory education: three hours a week would be set aside for training, one hour each for men and women as soldiers, citizens, and individuals (Bickerstetch 1944, Crang 2000). Furthermore, ABCA was increasingly interested in town planning issues as topics for discussion groups. In September 1942, *Current Affairs* published an issue on "Town Planning" by MARS group secretary Ralph Tubbs. In December, a member of the ABCA staff, Major Wakeford, met with Tyrwhitt to ask APRR to write something for *Current Affairs*. Tyrwhitt elatedly shared this news that same day over lunch at PEP with Forrester, Stead, and Pepler, who were on their way to the Board of Education to discuss funding for the School of Planning. [12/3/42] Tyrwhitt interacted with ABCA staff on APRR's pamphlet as she developed the curriculum for the correspondence course. Along the way she established the contacts and credibility with the Directorate of Army Education that convinced them to issue the course.

Curriculum Making as Collaborative Planning

It was with the correspondence course in mind that Tyrwhitt attended the Winter School of Sociology and Civics, organized by the Institute of Sociology (Le Play House) in early January 1943. She was particularly impressed by the lecture by Karl Mannheim, in which he argued that real curriculum making is collaborative planning:

> A curriculum is an example of real Planning, of genuine teamwork … . Or you may compare a real curriculum to a musical score written for a large orchestra. Just as it would be absurd to conceive of such music as a compromise between competing instruments, it is impossible to think of a curriculum except in terms of parts related to a clearly envisaged educational goal. (1944: 5)

Mannheim's talk may have inspired Tyrwhitt to think in a new way about teaching as well as curriculum design. He concluded: "in the democratically-planned society, the teachers will have to play one of the noblest parts. They can only achieve this by regarding their work as a serious responsibility to the community" (1944: 9).

A few weeks later, TPI gave formal recognition to APRR's School of Planning. Sadly, Dr. Stead had died unexpectedly shortly before then. Tyrwhitt would have to take the lead in the application for a Board of Education grant for the school, since Forrester had more and more on his hands, and Pepler was busy at the newly established Ministry of Town and Country Planning. [2/3/43] Then the school's advisory board asked Tyrwhitt to flesh out her summary curriculum for the correspondence course, which was organized in three terms—basic principles, tools, and methods—and fully develop the first. [2/28/43]

APRR and the School of Planning

Meanwhile, Tyrwhitt had completed her summary of the school's curriculum and the Association's work. This statement, published as a revision of APRR's Broadsheet No. 2 in April 1943, offers a definition of planning in line with Mannheim's notion of "planning for freedom," and incorporates the ecosystem concept of dynamic equilibrium:

> By "Planning" is implied the intelligent use by a free community of its environment, for the common good of its neighbours, its successors and itself. Such planning must grow freely and calls for careful, balanced thinking ahead and not for the rigid discipline of the blue print; once this balanced thought has been achieved, a delicate but constant adjustment to the needs of the environment can take place. (APRR 1943: 1)

APRR's statement presumed that "both planning and execution call for individual initiative tempered by group responsibility." APRR's sole task was "to ascertain and digest facts, and to present information and ideas in a form readily available." The role of a central planning authority is only to lay down "broad lines of policy and practical standards," leaving detailed planning to appropriate regional or local groups.

Tyrwhitt's synthesis articulated how the training of planners fit into the overall organization of APRR, and the relationship between SPRRD and Rowse's prewar school:

> As an integral part of its organisation the Association endeavours to train a nucleus of men and women in the technique of planning in its broadest sense, and in this respect ... has the full advantage of the experience gained by the School of Planning and Research for National Development during the six years up to the outbreak of war. A new body ... has been formed to carry on the work of the school, namely, the School of Planning and Research for Regional Development. ... Its immediate aim is to establish a correspondence course in the elements of planning suitable for those now serving with the armed forces and for prisoners of war. (APRR 1943: 1–2)

Correspondence Course: Part I, Background to Planning

Tyrwhitt's thinking about how to develop the first part of the correspondence course, titled "Background to Planning," was stimulated by an invitation she accepted in April 1943 to give an evening lecture course in the history of town planning at Regent Polytechnic, beginning the following week. She had one day to prepare her debut lecture—"a history of Europe from primitive man to 1500." She based this almost entirely on what she had learned from Rowse, but after rereading his recent papers—on a "grandiose" idea for the redistribution of population throughout the Empire—she became more cautious. "There's no doubt the man's a genius – but terrifying." She would have to turn to other sources for future lectures. "Its going to be the hell of a game tho' – preparing a lecture every week!" After writing her second lecture, though, Tyrwhitt felt "exalted not tired. I suppose it's the creative urge getting some satisfaction. Anyway I certainly don't regret giving up my week-ends and I'm learning a whole heap for myself." After a few lectures, Tyrwhitt could "see what 'gets' teachers!" this changed how she approached the correspondence course as well. [5/2-17/43]

But she was perhaps trying to do too much, though, and as a result not doing all of it well enough. "Work is very heavy just now – & I am pretty overdone really," she admitted on May 4th, and often found herself improvising. After a meeting at the Board of Education on May 21st she wrote: "Prof. T. was most scathing about my synopsis for the School's first term that I sweated over last week-end." In late May the School board agreed to form a team of three people who actually had experience as educators—Prof. Taylor, Lovett, and Lock—to help Tyrwhitt with the syllabus. The board also agreed to concentrate on that course, rather than continue to also try to fund the school.

It was "a momentous day" in June when The War Office and APRR agreed to start accepting students into the correspondence course in September or October. By mid-August, the ten lectures on "Background of Planning" were complete—"& all written by slap up authors." [8/17/43] The Foreword to the First Lecture explained: "this course … is addressed to men in the Forces who either intend to become Town Planners, or who wish to be able to take a responsible and informed part in the physical reconstruction of this country after the war."[1] The lesson itself was an introduction to the "Modern Concept of Planning" as pioneered by Geddes. The Geddessian approach to planning provided the guiding principles for the entire three-part course in its focus on "the need to be interdisciplinary, the use of the region as a planning unit, the necessity of a holistic approach, and the importance of economic and social factors." (Meller 1990: 323) After some delays the War Office *Correspondence Course on the Background of Planning* was printed in November. A few weeks later, a new chapter in Tyrwhitt's career began when the first students arrived by mail: "Quite a good mixed bunch. Rotten fun." [12/14/43]

---

1   "First Lecture, Foreword," working copy in author's personal collection.

INTRODUCTION TO COURSE

This third-part of the Course in Town Planning is intended to show the methods by which the student can employ the tools described in Part 2 according to the principles outlined in Part 1. The lessons in this part have increased in length as there are few current textbooks to which reference can be made. They will repay careful study in addition to that required to answer the actual questions set.

It will be noticed that Lesson I serves as an Introduction and has no questions. It is, however, necessary to read this lesson carefully before proceeding with the Course. It must again be studied at the end of the Course before answering the questions set to Lesson XI. Lesson XII is in the nature of an Appendix.

" A " students who successfully complete this course are eligible to attend Special Three Months' Completion Courses, which will be held in London and elsewhere as soon as the war is over. These will concentrate upon survey work, studio work and illustrated lectures on history and design—all subjects that cannot well be taught by correspondence. At the end of this Special Three Months' Course students will sit for examinations that will exempt them from the Final Examination of the Town Planning Institute.

CONTENTS

Part I.

		Exercise Numbers	Page Number
Lesson I. Introduction : Regional Integration, Part 1	.. ..	None	3
Lesson II. Regional Planning—Land Use	.. ..	(1) (2)	11
Lesson III. Regional Planning—Community Services	.. ..	(3) (4)	17
Lesson IV. Rural Planning	.. ..	(5) (6) (7)	23
Lesson V. Power in Relation to Regional Planning	.. ..	(8) (9) (10)	39
Lesson VI. Decentralisation	.. ..	(11) (12) (13)	53
Lesson VII. The Instruments of Transport and their Synthesis	.. .. ..	(14) (15) (16)	61

LESSON I.

REGIONAL INTEGRATION

Reading   Creative Demobilisation : E. A. Gutkind : Kegan Paul : 1943.
Culture of Cities : L. Mumford : Secker & Warling : 1940.
Evolution of Cities : P. Geddes : 1913.
*T.V.A.

American Publications—
American Regionalism : H. W. Odum and H. E. Moore.
National Resources Planning Board.
Regional Factors in National Planning 1935.
Pacific Northwest 1936.
Upper Rio Grande 1938.
Problems of a Changing Population 1938.

SYNOPSIS.

The Problem. Part I.

What is Regionalism—Regionalism and Provincialism—Why is integration necessary—What is to be integrated—Interdependence of social and economic integration—Regionalism as a readaption—Main factors in regionality.

The Approach. Part II.

Approach from the top and approach from the bottom—Growth from within—Delimitation and marginal areas—Functional and personal factors—Unity, diversity and uniformity—Interaction of town and country—Inter-regional balance and regional homogeneity—Systematic planning, not laissez-faire—Co-operation of the people—Education for regional consciousness.

Figure 7.1 **For the War Office Correspondence Course in Town Planning Tyrwhitt had to create lessons on topics on which few current textbooks were available**

1943 Postscript

The tide of war in Europe shifted definitively in favor of the allies during 1943, but the year ended on a sad note for Tyrwhitt; the War Office confirmed Cuthbert's death—he fell in the street fighting during the fall of Singapore in February 1942. Both Forrester's youngest brother and Tyrwhitt's old friend Stephen Bull had also been killed in action. On Christmas Tyrwhitt noted: "Of 18 first & second male cousins 8 have been killed; 3 are prisoners 5 still fighting, one still at school & only [one] has a civil job." The following excerpt from a poem Tyrwhitt wrote around this time demonstrates how these losses fueled her intensely personal sense of mission to plan a better post-war world:

> Their vivid lives have snapped, while we
> Who plod the treadmill endlessly
> Live on; grey drudges in the grey
> Hard-bitten towns of yesterday.
> Theirs was the future; ours to clear
> Away the dross of yesteryear.
> Till that the torch of their bright lives released from strife
> Should warm and quicken our chill plans to a new life.
> ...
> While it is clear that plans must be
> In touch with life or atrophy
> How arrogant it is to plan
> To be all things to everyman.
> Others live on and they must choose
> Such plans as they may wish to use
> Our task to polish each plan with such friendly care
> That many may clearly see the purpose mirrored there.[2]

Collective Leadership: Survey Before Plan

While working with Lock on the syllabus, Tyrwhitt continued to consider opportunities to partner with him to do survey work on a contract basis. After a strategy session on one potential client, she complained: "We shall obviously help with the job – but I think Max wants to keep the publicity etc. He knows I'm not over keen on this." [12/11/43] In early 1944, Tyrwhitt mobilized APRR resources to assist Lock on a social survey of Middlesbrough, as part of a larger project credited to the Max Lock Group. Tyrwhitt helped Lock organize and schedule the phases of the overall project and assembled a team of APRR researchers that

---

2   The poem titled "C.T. birthday II.IV.43 & J.F. brother killed" appears on a memoranda page in her 1943 diary.

included sociologist Ruth Glass, geographer AE Smailes, and herself. They lived together with other members of Lock's team in Middlesbrough, between April and October 1944, while conducting the field work (Glass 1948). Tyrwhitt mediated conflicts that arose between Glass, who had a baby and did not want to be away from home, and Lock. "Finally I've arranged that Ruth does the London end. I do all the visits to here & all the local contacts," she reported: "This way I think the peace may be kept." [4/3/44] She also collaborated on the analysis and mapping of the survey data.

It's important to recognize the essential contribution of such un-credited "collective leadership" to the practice of group work in general and the cooperative enterprise involved in the preparation of surveys in particular; Tyrwhitt specialized in supplying it and APRR began to promote this specialization (APRR 1944a).

Correspondence Course: Parts II and III

By March 1944 there were 55 students enrolled in the first term of the correspondence course, and the school advisory board directed Tyrwhitt to get the second term material edited at once. Writing the second and third parts—"Planning Factors" and "Planning Practice"— gave Tyrwhitt an opportunity to systematize the techniques she had devised for the analysis and presentation of survey data; in this way she integrated planning theory into a pedagogy for practice. Tyrwhitt completed her lecture on "Surveys for Planning" and worked on the "History of Town Planning" lesson while living and working in Middlesbrough in early April. By then the School of Planning was APRR's biggest job. Prof. Taylor forced discussion of that issue at APRR's April board meeting. "It had the effect of making Jim [Forrester] take the school seriously" Tyrwhitt observed. [4/13/44] Forrester's new focus on school affairs likely led to Tyrwhitt being named Director of Studies of the School—"despite myself," she declared, having apparently internalized the prevailing prejudice against women as leaders.

Tyrwhitt ran the correspondence course practically single-handedly from that point on. She finished editing the second part of the course while traveling to and from Middlesbrough on crowded trains, and sent it off to the War Office for printing in May1944. By then the course had 240 students. Approximately 1,600 enrolled altogether. [T.5]

APRR and SPRRD: Merger of Planning Research and Training

The end of May 1944 marked the beginning of a new phase in the life of APRR and its sister organization, SPRRD, as they became legally and financially connected, and moved into larger offices—nos. 34 and 35 Gordon Square—thanks once more to the backing of Lord Forrester. [T.3]

Tyrwhitt's personal and professional lives became even more intertwined when she moved into the apartment above the new offices. The impetus for Tyrwhitt's move was the marriage of her roommate Alison. In the swirl of events surrounding the wedding, Tyrwhitt had allowed herself to become swept up in a romance with Alison's brother. When this admittedly misguided affair ended, Tyrwhitt feared she might never again dare to dream the "dream of common womanhood, a man, a home, children, a garden rood."[3] Her feelings of having been "found unworthy" as a mate led her to lead a compensatory life: to do things for others "so that they should look on me with favor."[5/30/44] She wrote those lines as she prepared to move "above the shop." Her new apartment soon became an informal annex of APRR.

APRR announced in the revised version of Broadsheet No. 2 published in May 1944 that the work of the School of Planning "will be extended as rapidly as its financial resources allow," in anticipation of the British government's demobilization period education programs. It was time for the TPI Heads of Schools Committee to take action on the post war completion course for service personnel who had taken the War Office Correspondence Course in Town Planning, and who wanted to qualify for TPI membership. In the meantime Tyrwhitt was involved with TPI on several fronts, including designing a program for the annual summer school, which she completed on June 1, 1944. A week later, TPI asked Tyrwhitt if she would design the three-month post-war course—over the next weekend. "It's an honor – but a hell of a job." [6/8/44]. Two days earlier the Allies had launched their invasion at Normandy. Tyrwhitt happily recorded that Forrester arrived at APRR that evening "full of hope & plans for the future." Two weeks later he circulated a memo calling for consolidation of APRR and SPRRD administrations, and a fund raising campaign on behalf of the school.

In a parallel development, at its July meeting the TPI Heads of Schools Committee accepted Tyrwhitt's syllabus for the post war completion course. "That's a great triumph & I feel grand about it!" she exclaimed. [7/19/44] But the committee asked her to shorten it to conform to the length of university terms. This meant that any university-based planning program could offer the course in competition with SPRRD. Fortuitously, at the TPI summer school in mid-September Tyrwhitt agreed to fill in at the last minute for John Dower, who was to speak on the Education of the Planner. This gave her a platform to inform the professional community about the Correspondence Course in Town Planning and position SPRRD as a logical provider of the completion course.

A month later, Pepler chaired the first meeting with all of the key players involved in the Correspondence Course, including: the War Office; Ministries of Transport, Works and Building, Town and Country Planning, and Health; London and Liverpool Universities, and the four professional institutes represented on the Town Planning Joint Examination Board. There were now 525 enrolled students. Tyrwhitt reported on the status of plans for the post war completion course (APRR

---

3   Tyrwhitt poem, untitled, dated May 25, 1944, appears on a memoranda page in her diary for 1944.

1944b). However, as the year drew to a close she was still negotiating with TPI to recognize SPRRD's course. Pepler ultimately worked out an arrangement. [T.3] As Britain's new planning system became institutionalized, planning education would be restricted to university-based programs.

A Creative Synthesis

In the midst of all these activities, Tyrwhitt wrote an article in spring 1944 on "Town Planning" for the inaugural issue of *The Architect's Yearbook* (1945), articulating the synthesis of the Geddessian line of planning thought and modernist ideals that was being forged in the context of the work of APRR and SPRRD and their collaborators. She composed this statement in the context of her own growing involvement with the MARS group. Jane Drew, a MARS group member and a founding editor of *The Architect's Yearbook*, explained that the editorial board of that journal was committed to showing how the social-aesthetic ideals of the European modernists of the 1930s could be adapted to post-war conditions in Britain (1945: 5). In the realm of town planning, modernist ideals had been codified in the Athens Charter, which was written by Swiss architect Le Corbusier, based on discussions at the fourth CIAM Congress in 1933, and adapted for an English speaking audience by Spanish architect Jose Luis Sert in his book, *Can Our Cities Survive* (1942). "Town Planning" is Tyrwhitt's first contribution to that discourse.

Tyrwhitt (1945: 11, 13) acknowledged the limits of pre-war CIAM principles by presenting town planning as a discipline encompassing: The Region, The Neighborhood, Work, Food, Health, Education, Transport, Leisure and Holidays— not simply the four "urban functions" of Dwelling, Work, Recreation and Transportation stipulated in the Athens Charter. She followed Geddes in establishing the region as the basic unit for planning, stating: "towns of many sizes can be countenanced, provided ... that, at one level, they fit into the general framework of the region and, at the other level, they are suitably differentiated into coherent neighborhood units." Such a synoptic approach required an interdisciplinary team, but it didn't make any difference who was team leader. "The purpose of the team is to see the region always as a whole and, by pooling the individual knowledge ... of its members, to enable a balanced and dynamic development continually to take place." The team employs survey techniques to analyze the region in all its complexity, and provide the basis for a planning process that is both scientific and democratic. Widely distributed and clearly presented survey maps would provide citizens and members of parliament "a reasonable basis for informed criticism and judgment of ... planning proposals."

In making her case Tyrwhitt (1945: 13–16), invoked "the space-time scale of our generation [that] has been grandly set forth by Giedion and needs interpretation in all forms of physical planning,"—a reference to the already canonical *Space Time and Architecture* (1941) by CIAM general secretary Sigfried Giedion, which Tyrwhitt had been "enthusiastically introduced to by Bobby Carter a couple of years earlier." [T.7] Arid garden cities were not the solution: "The life of the future needs the two contrasts in scale expressed in the same plan: a sense of space, freedom of movement, scope for expression, together with closely knit neighborhood life." Humanistic mastery of our technical abilities—dramatically advanced by the war—depended on "intimate neighborhood life ... [that] breeds social consciousness and civic responsibility." This lesson was brought home by civil defense measures during air raids which demonstrated the value of "some form of common meeting-place" for neighbors.

Tyrwhitt celebrated the Peckham Health Centre as a model neighborhood environment where "positive health can ... be encouraged by the full and free development of the varied potentialities of each individual ... . within the pattern of the community" (1945: 23) In terms closer to utopian socialism than cooperative capitalism she wrote:

> Healthy people do not want to be organized, but they do want opportunities to do things together. It is on these lines that we can imagine our neighborhood environments of the future. Facilities that will be communally owned and communally run by the local people, with the doctors moving about as part of the social make-up of the whole. (1945: 23)

Tyrwhitt continued to elaborate her synthesis of Geddessian and modernist planning ideas in the years ahead, and inspired an influential cohort who returned from war service to take part in rebuilding blitzed and blighted areas of Britain, and the reconstruction of the post war world.

Part III
1945–1949

Chapter 8

Trans-Atlantic Shift

Reviving International Relations

With the approach of the end of the war, Tyrwhitt looked forward to leaving APRR, having fulfilled the duty for which she had been called into service. Inspired by memories of an idyllic life in Linwood, she decided to study forestry with the aim of connecting trees and town planning. Instead, she found herself among those re-establishing international relations that had been interrupted by the war. Early in 1945, while bombs and rockets continued to rain on southern England, the British Ministry of Information invited Tyrwhitt to undertake a lecture tour of Canada that spring to report on town planning for post-war Britain. The Ministry knew Tyrwhitt through her work with APRR and the correspondence course; she was an eminently suitable emissary: knowledgeable, poised, non-partisan *and* available to travel for months. Her tour was timed to coincide with the United Nations (UN) Conference on International Organization in San Francisco (April 25 to June 26). Jacob Crane (1892–1988), assistant to the Administrator of U.S. Housing and Home Finance Agency for international affairs, arranged for Tyrwhitt's tour to include US cities. Crane was also a Geddes disciple, and wanted to strengthen connections among like-minded planners. Tyrwhitt began her "Geddes job"— probably *Patrick Geddes in India* (1947)—while on the boat to Canada. [4/3/45]

Canada

The North American lecture tour proved to be a life changing experience for Tyrwhitt, opening new horizons and significantly extending her personal and professional networks. The contacts she established on this trip made it possible for her to launch her trans-Atlantic career despite her limited financial resources and restrictions on travel.

After arriving in Nova Scotia on April 7 1945, Tyrwhitt embarked on a hectic speaking schedule, traversing Canada by train in 27 days, making 21 stops in 17 different cities. She enjoyed the trip heading West, where the long journey between destinations was "restful and exciting at the same time." It was "grand to wake up in the morning beside a moving panorama of new countryside. Countryside is far too tame a word though – new horizons is more fitting." [4/18/45] In Edmonton, Alberta, the view from the airport observation tower offered a thrilling perspective on the world as a whole:

> Oil, gold & minerals were coming in from the North West territories &, in one of
> the rooms harsh discordant wireless messages were coming in from planes flying

anywhere west of Winnipeg, north of Minneapolis & south of Whitehorse near the Alaska border. One really believed in polar routes & global communications in this place ... . Everywhere seemed suddenly near at hand, turn to the left for Moscow, right for Glasgow. [4/21/45]

Tyrwhitt's lecture tour was co-sponsored by the Royal Architectural Institute of Canada. Town planning as a profession distinct from architecture was still in a formative stage there; there weren't any university training programs. Tyrwhitt knew two key members of the nascent planning community from her student days: John Bland, who graduated from the AA School of Planning in 1937, was now director of the School of Architecture at McGill University in Montreal; and Humphrey Carver, her "first hero at the AA" when she was 19, now based in Ottawa, was a leader in the field of social housing and community planning. [5/6/45] Tyrwhitt concluded that the time was ripe to establish "at least one course in town planning—on a post-graduate basis" in Canada to train students for this new profession for which there would be growing demand in the post-war years. [5/20/45] Tyrwhitt buildt on the connections she made on this trip to lay the groundwork for such a program in Toronto in 1953–4.

The specifics of Tyrwhitt's Canadian talks varied according to her audience and venue—civic groups, "ladies' tea," or technicians—but, "the main job was to get them to see that a survey must be followed by a master plan & that a master plan can only be prepared by an outside expert [training & objectivity essential]." [4/12/45]

United States

On her way to Boston, Tyrwhitt admitted some trepidation about what she was about to face. "[Americans] are, in most planning matters, well ahead of us as far as I can find out—so that one cannot get away with the sort of generalities" she used in Canada. [5/17/45] Upon her arrival she declared: "Well, if I didn't get my head turned in Canada I certainly will here!" She declared Boston to be "the first real <u>town</u>" she had been to in North America. "It's not only the congestion in the centre that gets me, but the lavishness of everything. The food is something to dream about," she wrote. (In Britain the watchword was still austerity.) Moreover, both the Massachusetts Institute of Technology (MIT) and Harvard University were there, "and there's a City Planning Board and a State Planning Board and a Metropolitan Plan," she reported, "somewhat of a galaxy." The first afternoon she met at Harvard with 30 of the stars in that galaxy, including Catherine Bauer Wurster, then a lecturer in housing at Harvard's Graduate School of Design (GSD), with whom she struck up a friendship, Katherine McNamara, GSD librarian, Martin Wagner, former Berlin city planner and now Professor of City Planning at GSD, Talcott Parsons, chair of the Sociology Department, and renowned agriculturalist J.D. Black. GSD Dean Joseph Hudnut hosted "a formal gala lunch ... with all the names one's ever heard of," including Walter Gropius, then head of the architecture department, and Holmes

Perkins, who was then about to offer a class in regional planning. "I was more than a shade over-awed," she acknowledged. [5/18-21/45]

Initially she was anxious: "I had no idea what I was supposed to be up to in the States nor what they were interested in." That first day at Harvard she fielded "a range of questions from housing management, siting of industry, interim development, Uthwatt report, agricultural planning, local government committee, and so on," and she "just let rip – and apparently got away with it." [5/18/45] The rest of her US tour followed this pattern: in each city she was feted and introduced to civic leaders and top professionals and academics in planning and related fields. The more leisurely pace of this leg of her trip—in 32 days she visited Boston, New York, Cincinnati, Chicago, Detroit, Knoxville (Tennessee Valley Authority headquarters), and Washington, D.C—allowed for more time to get to know the people and places she visited. Many "housers" and planners already knew of Tyrwhitt based on her work during the war.1

In New York, where she celebrated her 40th birthday, she met her American sister-in-law, Delia Tyrwhitt, Cuthbert's widow, for the first time since 1937. Tyrwhitt was relieved to find Delia—who had escaped from Singapore just before it fell to the Japanese—"quiet & demure & gay & courageous & utterly delightful." This meeting marked the beginning of their life–long friendship. Delia was an accomplished photographer, fluent in Chinese, and had "an intimate knowledge of life and personalities pretty well all over the East." [5/24/45] She was wealthy but a little lost, and turned to Tyrwhitt as a guiding light.2

Tyrwhitt had a glittering introduction to the modernist elite in New York at a party hosted by the British Information Service. This was where she first met Swiss art historian and CIAM secretary Siegfried Giedion. Other guests included William Lescaze, Austrian graphic designer and former Bauhaus teacher Herbert Bayer, Philip Goodwin, a board member and architect of the Museum of Modern Art (MoMA), and one of the designers featured in that exhibit, Russian-born English architect Serge Chermayeff, a member of the MARS group until he emigrated to the US in 1940. She felt comfortable among this crowd: "We're of approximately the same outlook aesthetically & politically & had all had a reasonable acquaintance with pre-war Europe. We were realists — realistic idealists?—aware of the hopefulness of the next few years but acutely conscious of the dangers." [5/27/45]

Tyrwhitt's membership in the MARS group (as of February 1945) ensured her a "warm welcome from CIAM exiles in the US." [T.7] In Chicago she met former Bauhaus teacher Lazlo Moholy-Nagy. Tyrwhitt was impressed by Moholy-Nagy's work at his new school, the Institute of Design, where he was adapting the Bauhaus approach. "He is training designers who are not afraid of the machine, but learn to use it as a … collaborator," she enthused. Tyrwhitt also shared Moholy-Nagy's vision of "happy and organic cities" where "inhabitants have the experience of being amidst gardens and vegetation daily" (1945: 11). She walked back to her hotel "on air." [6/6/45]

---

1 Marjorie Meyerson, personal communication with author 8/4/08.
2 Daniel Huws, personal communication with author 7/30/11.

In Chicago, Tyrwhitt also spent an interesting and informative time with Walter Blucher, president of the American Society of Planning Officials (ASPO), which was housed with a number of other like-minded organizations—notably, the Public Administration Clearing House (PACH)— in a building known by its address, "1313." Those organizations shared the aim of promoting the exchange of information and fostering cooperation between groups concerning public administration, from the local to the international level (Saunier 2001: 389). Visiting Moholy-Nagy's school and meeting Blucher and his colleagues at 1313 inspired Tyrwhitt to reflect on her work from an international perspective:

> I am determined to get the School course going at Gordon Square; and then I would welcome an excuse to hand [it] over to someone back from the Forces. The Association itself may gradually cease to be necessary as the whole country gets into its stride, or it may again become a research annex to the school, or it may become straightforwardly commercial & do just contract work or, with some good background grant ... it may grow along its present lines. But I think its character is bound to change. The finest thing it could do would be to co-ordinate physical planning thought in tackling Europe. [6/9/45]

At the end of her trip, Tyrwhitt lunched with Moholy-Nagy and Giedion in New York. A quarter of a century later she reflected on the profound significance for her of those meetings:

> Although I had an architectural background, my mind was almost wholly occupied with the social and economic aspects of the problem, and the world of art was deliberately disregarded. My contact with Moholy Nagy in Chicago changed all that, and when I met him again in New York with Giedion, I experienced a sort of conversion, somewhat similar to suddenly "getting religion." My eyes were opened. I continued my former work but with a different viewpoint. [T.7]

APRR: Carving a New Path

Upon her return to London, Tyrwhitt organized a conference on "Human Needs in Planning: the Contribution of Social Studies to Architecture and Planning," which APRR convened in cooperation with the RIBA Architectural Science Board and the Institute of Sociology in late November 1945. The purpose of the conference was to enable organizations and individuals engaged in social surveys relating to physical planning "to compare and discuss their methods and findings" (Human 1946: 126). It was a logical follow-up to the meeting Tyrwhitt had organized with Carter in 1942 on regional survey techniques. It also gave APRR an opportunity to spotlight its surveys of Middlesbrough and Bethnal Green, largely the work of Ruth Glass, who was an important part of APRR's "composite mind." Tyrwhitt hoped those surveys would "at last really reduce the amount of rubbish talked about 'neighbourhood units,' 'sociology applied to planning' and so on." [4/6/45] Glass reported on the difficulties of achieving social diversity within a single community: different social groups live in separate and distinct geographical areas and have neither the need nor opportunity to share institutions (Human 1946: 127).

For a meeting of experts the Human Needs in Planning conference attracted significant media attention, as it addressed a key assumption underlying the new planning legislation: the nature and availability of useful social data on which to base planning and policy decisions. The conference confirmed: "the science of social studies is necessary in a nation organized on modern communal lines, but ... we have still a long way to go in meeting human needs and that there are subtleties that many of us have hitherto not realized" (Human 1946: 126, 128).

Information Service

Perhaps inspired by her visit to "1313," in the fall of 1945 Tyrwhitt oversaw the launch of an expanded, subscription-based version of APRR's Information Service. The Association (1945b: 775) announced in the journal *Nature* that subscribers would be offered, "in addition to copies of all the Association's broadsheets and reports and the services of the library and the information bureau, a bimonthly bulletin containing a report of work being undertaken by the Association as well as a resume of matters of interest in the planning world and pre-publication news of books produced by the Association." Such a service made good business sense, as sale of publications had become APRR's only "money raisers." [7/4/44] There was a need for this service in Britain, as APRR explained in a brochure in 1946:

> In this country the planner is continually frustrated because statistical data is unavailable. Accurate information about home and world trends of production and consumption, about population trends and wages, about the movement of capital, price levels and so on are absolutely necessary; ... . Yet, there is no counterpart in this country to the excellently organized and up-to-the-minute statistical service of the United States. Given time and opportunity APRR hopes to give assistance to those who need current research in particular fields.[3]

The visualization of data was one of the services featured in APRR's 1946 brochure, which noted that in addition to the "vivid presentation of information by Maps," APRR "have introduced this graphic method to illustrate statistics on such subjects as Extractive Industries, Trade and Transport, the Cost of Living, the Old and the Young, the Prevention and Treatment of Sickness, and many others."[4]

---

3 The Association for Planning, brochure, London, (1946: 8) in author's personal collection.
4 The Association for Planning, brochure, London, (1946: 1) in author's personal collection.

APRR also promoted its compilation services: "the collection and presentation of facts in text, maps, diagrams, graphs and photographs for industrial and other organizations, and for authorities and individuals." A good example is *Housing Digest: An Analysis of Housing Reports 1941–1945* (1946a) which Tyrwhitt prepared for APRR on behalf of the Electrical Association for Women. This category of service also included the submission of evidence. In February 1946, for example, APRR submitted evidence to the Committee on New Towns, under the direction of Lord Reith, which had been convened by the Ministry of Town and Country Planning. Tyrwhitt notably warned that existing planning principles, such as the concepts of the Neighborhood Unit and the Green Belt, which were derived from surveys of older, larger towns, "are not necessarily applicable to small towns in general or to New Towns in particular."

In 1945 APRR also introduced a series of popular booklets called "Survey before Plan." APRR had long used the slogan "Survey before Plan" to align its work with Geddes's ideas. The publisher, Lund Humphries, had approached Tyrwhitt in the summer of 1943 about APRR preparing such a series on planning surveys. Her immediate reaction was: "we could and should." [8/5/43] Tyrwhitt was already working with Prof. Taylor on writing a similar pamphlet for *Current Affairs*. She revised Taylor's draft to serve as an introduction to the series. Subsequent booklets were to be edited volumes, "presenting in a less technical manner much of the material contained in APRR's broadsheets" (Taylor 1945: 1). Ultimately, only three were published.[5] Regardless, Tyrwhitt's role in conceptualizing the series as a vehicle for public education, and getting the first booklet— lavishly illustrated with plans, maps, diagrams, and photos—produced is clear, yet characteristically anonymous.

Tyrwhitt was particularly proud of APRR's library; she assembled most of the collection from her own books, and she hired librarian Ellen Schoendorf in late 1944 to organize and catalogue the collection. Under Tyrwhitt's supervision, Schoendorf adapted a version of the Universal Decimal Classification (UDC) system devised by the Belgian documentalist Paul Otlet (1866–1944) to suit APRR's work. One advantage of UDC over the Dewey decimal system was that it facilitated cross-referencing as required in a comprehensive approach to planning (Rayward 1994). Tyrwhitt would have known of Geddes's collaboration with Otlet "to find new ways to organize and disseminate knowledge on a global level" (van den Heuvel 2008: 128). Tyrwhitt shared Geddes's taxonometric bent, and loved working in libraries as much as she loved gardening. APRR's classification system emphasized the main topics of physical planning and related subjects (including documentation and library science) from the point of view of the planner (1950: 577). Thanks to Tyrwhitt APRR could boast in 1946 that its library's "collection is unique in its international range and cannot be matched in London."[6]

---

5   In addition to Taylor's *Planning Prospect* (no. 1) the Survey before Plan series includes E. W. Willis (ed.), *The Hub of the House* (no. 2); and M. George (ed.), *Wealth from Waste* (no. 3).

6   *The Association for Planning*, brochure, London, (1946: 2) in author's personal collection.

In conjunction with its growing collection, APRR launched a monthly Reference Sheet as a supplement to the Information Bulletin, listing recent library acquisitions and including brief articles and the Planners Digest, an annotated bibliography of selected references. The first, issued in September 1947, introduced the APRR Classification System. In this way, Tyrwhitt directed APRR to systematize the "broader conception of planning" called for in the Town and Country Planning Act of 1947 and to disseminate information relevant to this field to contribute to post-war reconstruction.

School of Planning: Completion Course

One practical reason for the development of APRR's library was to support the School of Planning's Special Three Months' Completion Course for those who had successfully completed the Town Planning Correspondence Course. Classes took place in No. 35 Gordon Square, despite a large hole in the drafting room, a souvenir of the blitz. Tyrwhitt based the curriculum on a lecture course given by Walter Blucher of ASPO; [6/5/45] and later elaborated:

> The three month's courses were designed somewhat on the basis of the courses in the old School of Planning – that is to say the students were subjected to a number of different individual lecturers … . But the need to compress so much in so little time meant that the course as a whole had to be drastically rigid and there was scarcely any time at all for general discussion. The feeling of bewilderment engendered by the first School of Planning courses was lessened to some extent by a strict unity of subject: in the first month all the lectures and studio work were concentrated on survey techniques, in the second upon the components of the plan, in the third upon the implementation of planning. That meant that all the law lectures, for example, came in a heap during the last month. As a system this seemed to work out very well, and I wish it were possible to operate under University conditions, but it must be owned that those three month courses were pretty indigestible – and very grueling both for staff and students. [T.3]

There was such demand for the three-month completion course that it ran for seven consecutive sessions, ending in December 1947. In addition, Tyrwhitt ran an Overseas Correspondence Course through the school beginning in autumn 1946.

Educational Philosophy

Tyrwhitt described her understanding of what students of planning ought to be taught and how in "Training the Planner," published in the reference book, *Planning and Reconstruction* in 1946. This article shows how she was melding the new viewpoint she had gained from her encounter with Moholy–Nagy onto her fundamental

Geddessian perspective. "A planner must be able to see the social life of a town and its physical pattern as one related whole," Tyrwhitt stated (1946a: 210–211):

> He must also know the effect of any change in one part of the town upon the life of the whole. Not only must he know this, but he must be able to anticipate how the town will grow or alter ... and he must, as importantly, know [peoples'] requirements. *Finally, a plan is a design and the planner must be a designer; ... the creative artist who not only sees what is in terms of what could be, but has the power to set this down in such a manner that his vision is shared and understood by others.* [emphasis added]

The planner's synoptic vision "must be the outcome" of postgraduate training. Undergraduate education could equally well be in four basic fields: the technical sciences; the social sciences; the physical sciences; or the graphic arts. Since one individual is unlikely to be fully competent in all four fields, the key is "the principle of group work." Ideally all four basic areas of expertise would be represented on a planning team, and training in planning would enable team members to appreciate the contribution of the others and collaborate—in other words form a "composite mind" (1946a: 11–12).

Significantly, Tyrwhitt recommended separating "the detailed study of methods of operating the Town Planning Acts from a general course in planning." This would shorten the course and make it "more attractive to members of varied professions whose cooperation is essential for the creation of integrated planning schemes." She presented the SPRRD's Completion Course as a model for a program that would qualify graduates "as people who had learnt something of a composite mind." The war had shown how effective the multi-disciplinary team could be. "Our chief need in training for planning today would seem to be to find an efficient and attractive method of imparting knowledge of planning to the potential collaborators in planning teams rather than to concentrate only upon training the administrative planning official (1946a: 212–213)."

Rowse's Return

Rowse returned from military service in autumn 1946. The APRR Information Bulletin for December 1946/January 1947 announced that Tyrwhitt would step down as director of studies for the School of Planning at the end of the Completion Courses. Rowse would assume those duties, but the bland notice belied the fact that Rowse would not *resume* his former position; he would be taking over a *new* school, one that Tyrwhitt had started, and a support network, including a faculty, which she had assembled and nurtured.

The recollections of students confirm that Tyrwhitt was the guiding spirit of the School of Planning. Architect Eduard Sekler, who arrived at the school in 1946 as a British Council scholar from Vienna, was surprised to discover that the "good looking, very dynamic young woman" who interviewed him turned out to be the "head of the whole institution." Sekler had never met anyone like her: "very British and somewhat awe-inspiring in her organizational efficiency. With infinite attention to detail she takes care not only of the smooth running of our classes but also of the individual welfare of each student." Sekler was "like most of the others ... completely won over by JT's infectious enthusiasm. One begins to sense that something of the spirit of William Morris is alive in this perpetually overworked teacher/administrator." [E.5]

Tyrwhitt worked with Rowse to organize a one-year full-time, postgraduate Diploma Course in Town and Country Planning, to open in September 1947. She later recalled viewing this as a temporary measure: "Until the University of London could get going again on a fulltime day course, it seemed the duty of the School of Planning to fill the need." [T.3] The Town and Country Planning Act of 1947 required all local planning authorities to prepare a comprehensive development plan based on a detailed survey, which, Tyrwhitt (1950a: xv) observed, "made the training of planner in the new techniques an urgent necessity."

Landscape Design: Training and Publications

Concurrent with the completion course, Tyrwhitt organized a postgraduate evening course on landscape design which she also ran through the School of Planning. She had suggested the idea for such an "emergency" course at an ILA meeting in October 1944. [There was only one pre-war diploma course in Britain in landscape architecture (Powers 2002).] Tyrwhitt developed the curriculum with Brenda Colvin, who taught it along with Brian Hackett. [11/8/44] This was the first evening course in Britain to be held in landscape design.[7] ILA conducted the final examinations, a task that went hand in hand with the establishment of professional qualification standards within the Institute in 1946. Significantly, this course paved the way for the new program in landscape architecture that was started under Hackett's direction in 1948 within the department of town planning at the University of London.

Moreover, a book project that Tyrwhitt had concocted with Colvin came to fruition during this period: *Trees for Town and Country* (1947a). They began work on this project in 1943; Lund Humphries agreed to publish the book in 1944. [1/3/44] Tyrwhitt and Colvin, along with illustrator S.R. Badmin, are credited as having prepared the book for APRR. Tyrwhitt's contribution appears to have been to serve, in addition to conceptualizing the project, as the overall organizer of the book, coordinating contributions of text and graphics, and overseeing the details of production with the publisher.

Trees for Town and Country (1947a: 5–7) described and illustrated 60 trees suited to Britain, profiling each tree's habitat, visual effects, and potential uses, with particular regard to anticipated post-war development. Notably, trees were recommended for use in towns considering "large new areas of open space ...

---

7   *The Association for Planning*, brochure, London, (1946: 9) in author's personal collection.

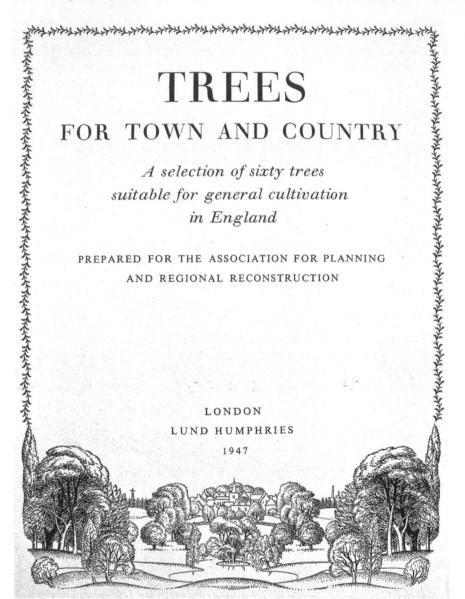

Figure 8.1   Trees were symbolic of the spirit of optimism that inspired post-war reconstruction in Britain: the building of a new, modern, healthy environment

under official Town Planning Schemes and the Education Act" [of 1944]); village greens for "the forthcoming big programme of rural housing"; and woodlands, as "large schemes of afforestation are afoot and the problem of relating these to the surrounding landscape is always one that excites controversy." *Trees for Town and Country* proved to be very popular; a second edition was published only two years after the first. The book also had lasting value. One British garden authority, writing in 1992, declared it "the best book on tree size and suitability" (in Gibson 2011: 99). Another praised it as "finely produced ... with concise information which is well conveyed. Trees were almost a symbol of the spirit of reconstruction" (Powers 2000: 59–60). Through this book and evening course, Tyrwhitt made a small but substantial contribution to the evolution of landscape architecture as a field of study in Britain.

While they were working together on the tree book, Tyrwhitt and Colvin were also active as members of the WFGA council committee on prospects for women in agriculture or horticulture (Twinch 1990). The committee's aim was to provide career advice to students at schools such as Swanley Horticultural College and decommissioned Land Girls who wanted to stay on the land. As landscape architects, women could engage with many aspects of post-war reconstruction, including planning of new towns, industrial location, and design of circulation systems, as well as regional planning around service farms and fresh food distribution systems—which was Tyrwhitt's pet concept.

Tyrwhitt's WFGA council and committee work appears to have given rise to her book *Planning and the Countryside* (1946b), a compilation of material prepared by others for APRR. Again, Tyrwhitt's contribution lay in her initiative, editing, and steering the book through production. *Planning and the Countryside* was aimed at a popular audience; Tyrwhitt intended it to build a case to counter the "more vociferous, more influential and more wealthy sections of the community who continually demand the best land for purposes that ...[are] of less importance in the long run, to the nation as a whole." The national interest would be best served by revitalizing the agricultural economy: "A healthy mixed farming tradition close to and surrounding the centres of population means healthy land and healthy people" (1946b: 1–2, 8).

The bulk of *Planning and the Countryside* describes a system and explains the benefits of service agriculture and what that entailed in Britain. Many of Tyrwhitt's proposals remain relevant to current discourses on sustainable community design, including: land use planning to ensure a local supply of fresh food; organizing health services to promote *wellbeing* not merely prevent or treat disease; ecological sewage systems; redeveloping existing towns prior to building new towns; and providing infrastructure to support village social and economic life (Gilg 1978: 353). However *Planning and the Countryside* confirmed that Tyrwhitt's regionalism in 1941–1946 was focused on the vitality of rural life rather than the dynamism of the city in the region.

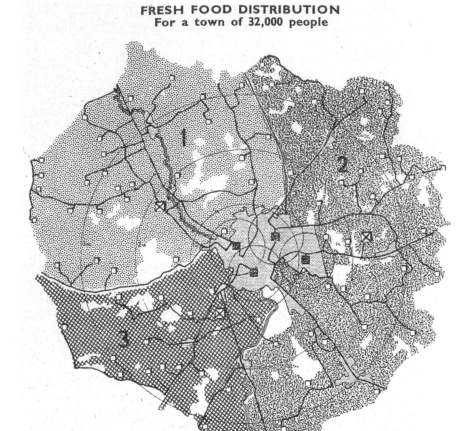

FRESH FOOD DISTRIBUTION
For a town of 32,000 people

FRESH FOOD CENTRES   **CONTROL FARMS**   **ORDINARY FARMS**
TOWNS AND VILLAGES   **WOODLAND, GOLF COURSES, ETC.**

The division into 3 sections is shown by varying tints.

Figure 8.2   **Tyrwhitt promoted ecological planning based on fresh food distribution systems**

APRR and SPRRD: Setback

APRR and the School of Planning suffered a serious setback when Lord Forrester was gravely injured in an accident in May 1947. He never returned to his previous level of involvement in APRR or the School, and they never recovered from the loss of his leadership and financial support. The School nearly closed in summer 1948. Since TPI would not recognize the new course in regional planning that Rowse wanted to develop—that would focus on world regions—the Ministry of Education cut its support for the School. The new grant "will be barely sufficient to keep the School afloat, and certainly not enough to support the Library," she wrote to architect Albert Mayer—whom she had met that year in New York—seeking help in finding a funding source outside of cash-strapped Britain. "There is no doubt that the present Library service could easily be developed into a valuable Planning Documentation Centre of international significance, if it could but be ensured of a basic income." [T.20] Such support did not materialize, but when the UN established a reference center within the Housing and Town Planning section of the Department of Social Affairs in 1949, APRR librarian Ellen Schoendorff was hired to run it. [B.1] The value of Tyrwhitt's work developing the APRR Library was not lost; it served as a training ground for the new international initiative.

Likewise many of Tyrwhitt's ex-soldier students went on to positions of leadership in the new planning agencies, ministries, and academic departments then being created in Britain. Summarizing her career in 1954, Tyrwhitt affirmed: "this is the work [she] likes to think she may be remembered for. It qualified 170 men for active service in town planning and got them into the field at the time the British Town Planning Act of 1947 needed them." [E.1] Her students also went to work in the Crown Colonies and in the Dominions, where they applied her lessons in seminal projects. All together they constituted a global "knowledge network" that facilitated the cross-fertilization and evolution of Geddesian modern planning ideas and practices.

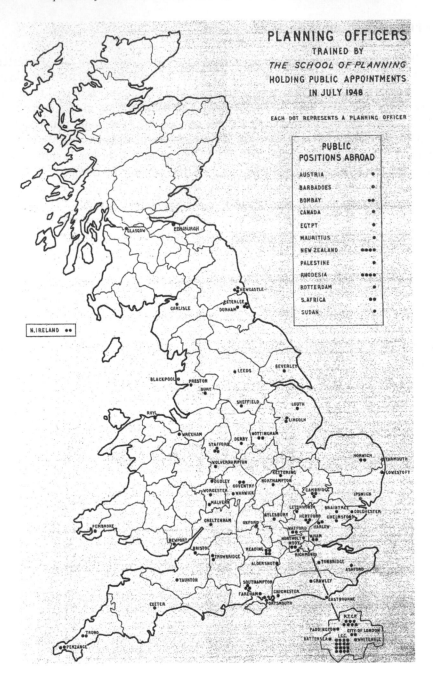

Figure 8.3   Former soldiers trained by Tyrwhitt assumed leadership roles throughout Britain, as well as in the Crown Colonies and in the Dominions

Chapter 9

Geddes as a Guide

Scientific Humanism and International Cooperation

There was a moment of optimism in the immediate post-war years, before the constraints of the cold war set in, when idealistic architects, planners, artists, scientists, and others believed they might actually realize the visions of a better world they had imagined during the war. The philosophy of Scientific Humanism pervaded the overlapping networks of these progressive reformers that converged in London and Paris in 1946 to renew their ties in the context of the new UN organizations then being established. This was made explicit by PEP founding member Julian Huxley, who was named Secretary of the Preparatory Commission for the United Nations Educational, Scientific and Cultural Organization (UNESCO) in 1945; he believed "the general philosophy of UNESCO should ... be scientific world humanism, global in extent and evolutionary in background." In Huxley's view, UNESCO's task was to "help in the emergence of a single world culture;" this could be gradually achieved by "securing the co-operation of peoples and nations and individuals representing different ideologies, on specific common tasks," such as a UNESCO Institute for Home and Community Planning (1946a: 6, 61; 1946b: 22).

Tyrwhitt soon was immersed in these networks and their visions. Huxley had recruited her old friend Bobby Carter as head of the library section of the UNESCO Preparatory Commission, which moved its operations from London to Paris in September 1946. Tyrwhitt caught up with Carter while she was in Paris that September to represent APRR at the Congrès Technique International in Paris. [T.21] (She spoke on "Factors influencing the layout of New Towns"—a topic of great interest as the British government had announced its intention to build twenty new towns over the next ten years.) Tyrwhitt's own work developing APRR's Information Service was connected to Carter's proposal to the Preparatory Commission—which was adopted as the program of UNESCO for 1947—which included as initial steps: "to set up a clearing house for exchange and distribution of books;" and "to give attention to bibliography" techniques (United 1947: 132). In July 1946, Tyrwhitt had proposed the idea to Catherine Bauer of establishing a Book Exchange arrangement, to support APRR's aim to "supply research workers in the planning world with up-to-date data on the planning aspects of *all* countries." Bauer replied immediately agreeing to the idea, adding that she looked forward to hearing the first-hand news of British planning from Lewis Mumford (1895–1990). [B.2] Tyrwhitt met Mumford when he visited Britain in June 1946, sponsored by the Institute of Sociology/LePlay House to participate in the Institute's first post-war International Conference of Social Work.

Immediately after the Paris congress, Tyrwhitt participated in a MARS group meeting with CIAM president Cornelus van Eesteren and general secretary Sigfried Giedion to discuss potential themes for the group's first post-war congress, CIAM 6, scheduled for September 1947; Tyrwhitt was on the congress organizing committee. Giedion and van Eesteren also were in London to forge an alliance between CIAM and the International Union of Architects (UIA), which was holding its first post-war conference at that time. Both CIAM and UIA hoped to establish a relationship with the UN through UNESCO, as non-governmental voluntary organizations.

Giedion was drawn to Tyrwhitt, as he had been at their first meeting in New York in May 1945. Since then, Giedion had returned to his wife and their two children in Zurich after having been stranded in the US for three and a half years due to the war. He was about to assume his first full time professorship at the Swiss Institute of Technology (ETH) (Georgiadis: 1997). Independently wealthy, Giedion could afford to travel to conduct his mission to revive CIAM. Over the next year Tyrwhitt and Giedion deepened their friendship as their paths crossed repeatedly in the context of planning for CIAM 6.

International Federation of Housing and Town Planning

A few days after meeting with Giedion, Tyrwhitt continued her conversation with Bauer, who arrived in London to attend the first post-war congress of IFHTP. Under Pepler's direction, the Federation had survived the war years based in London, and was holding its congress in Hastings in October 1946. To ensure that APRR would have a significant presence there, Tyrwhitt organized a parallel exhibition on "Plans of the Cities of Europe"—a reference to Geddes's "Cities and Town Planning Exhibition," which Pepler had helped hang at the 1913 International Exhibition in Ghent— and an associated one-day conference. In addressing whether and how European nations might take advantage of the opportunity provided by war damage to re-plan and rebuild urban centers, the congress focused attention on Britain's policy of planned decentralization via New Towns (Osborn: 1946). APRR's exhibition and conference papers (which were reprinted in the *Journal of the Town Planning Institute*, June 1947) provided an historical perspective on the New Town strategy.

Just as Geddes's exhibit displayed the evolution of settlements from cave dwellings to modern cities, Tyrwhitt organized APRR's exhibit in a chronological sequence: the Middle Ages, the Early Renaissance, Grand Renaissance, Black Cities of the 19th Century, and Green Cities of the 20th Century. A guide to the exhibit showed a typical plan from each of those "main town-building periods" together with "a contemporary description of the ideal town written by an observer of those times" (APRR 1946b: 1). Those were paired with "a short comment from a modern authority expressing present-day opinion on the results." For three out of five of these authorities, Tyrwhitt turned to quotations from her newest mentors: Giedion, Bauer, and Mumford. APRR's exhibit modernized Geddes's model, though, by incorporating aerial survey techniques, developed during the war, to enhance the display of conventional maps, plans and renderings. Likewise, the cover of the guidebook updated the bird's eye rendering that Geddes typically used to portray a town in its regional setting, using instead a view from space overlooking the cities of Europe (foregrounding England and Scandinavia) in a global regional setting encompassing north-western Eurasia, northern Africa, and, on the horizon, the Middle East.

APRR joined over one 1000 delegates from 23 countries at the IFHTP congress who adopted a resolution to "urge the Economic and Social Council of the United Nations to establish promptly within its own framework a unit to deal specifically with the international problems of housing and planned reconstruction" (Ihlder 1946: 13) Shortly thereafter, at UNESCO's first General Session, the program commission adopted the resolution to set up a section on home and community planning with the long-term view of establishing a semi-autonomous Institute (United 1947: 132).

Geddes in India: Bioregionalism

At this time Tyrwhitt also completed work on a complementary project: editing writings by Patrick Geddes, drawn from the town planning reports he made for Indian cities between 1915 and 1919. Geddes had done the majority of his actual planning work in India, but this material had not been previously available in Britain. Tyrwhitt's friend, Dr. H.V. Lanchester—who worked with Geddes in India (and helped replace the Cities and Town Planning Exhibition when the original sank en route to India)—had made an initial selection of material. In October 1944, Tyrwhitt persuaded Lund Humphries to publish the book. [10/23/44] Tyrwhitt agreed to edit the texts and find appropriate illustrations. Arthur Geddes assisted her. Tyrwhitt also enlisted Mumford's support; he wrote the Introduction to the edited collection, entitled *Patrick Geddes in India* (1947a) in September 1946. Tyrwhitt often turned to Mumford for guidance during the coming years as she continued to curate and amplify Geddes's sparse writings to reach new audiences.

Tyrwhitt completed work on *Patrick Geddes in India* in February 1947, and APRR (1946/47) announced its anticipated publication before the summer; this would coincide with India's independence from the British Empire. Tyrwhitt's intent was to demonstrate the relevance of Geddessian ideas to the current *worldwide* task of urban reconstruction and the realization of a new world order based on cooperation that UN and UNESCO would help foster. In an editor's note, Tyrwhitt (1947a: 6) explained that her task was "to choose passages that clearly illustrated the practical application of those town planning principles for which Patrick Geddes stood." Those principles included: "diagnosis before treatment," that is, survey before plan; and "conservative surgery," that is, rehabilitation rather than removal, an adaptive process that respected local tradition as a living, evolving, resource. In addition, the extracts highlighted Geddes's concept of "bio-regionalism:" "Environment and organism, place and people, are inseparable." That is "what makes this book particularly apt and timely for the days ahead," Mumford declared in his introduction. "In short, one cannot appreciate Geddes's regionalism unless one also appreciates

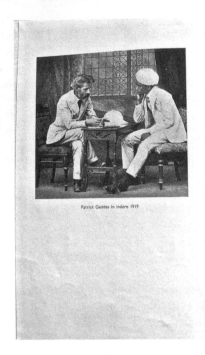

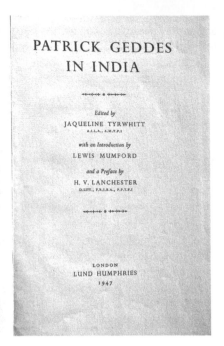

Figure 9.1   Tyrwhitt's edited collection of Geddes's writings presented material that was not previously available outside of India

his internationalism, his universalism ... What he says about India has a lesson for other lands." Geddes's thoughts on political decentralization, civic responsibility, voluntary cooperation, and personal development, "sound an even fresher and saner note today," Mumford affirmed (1947: 8–9).

Moreover, through Geddes's words, Tyrwhitt urged Westerners to learn, as Geddes did, from the "wisdom of the East," how to look at life as a whole (1947a: 26):

> [T]he cause of the frequent aesthetic failure of our results ... *is due to the lack of harmony between the advancing phases of western 'science'.* Each of the various specialists remains too closely concentrated upon his single specialism ... . In the east, on the other hand, it has been the glory of the historic sages and ancient rulers to concentrate their minds and efforts upon life as a whole. As a result, *civic beauty in India has existed at all levels*, from humble homes and simple shrines to palaces magnificent and temples sublime. *In city planning then, we must constantly keep in view the whole city, old and new alike in all its aspects and at all its levels.* [emphasis added]

Tyrwhitt underscored this lesson through the use of photographs taken during the war, as well as aerial views, to illustrate the aesthetic qualities Geddes admired in

India's built environment: "The transition in an Indian city from narrow lanes and earthen dwellings to small streets, great streets, and buildings of high importance and architectural beauty, form an inseparably interwoven structure."

Mumford promoted Geddes's ideas through his writings, notably in *The Culture of Cities* (1938), but with *Patrick Geddes in India,* Tyrwhitt made Geddes's own words comprehensible to a broad audience. That wasn't easy to do. Tyrwhitt (1947a: 6) not only distilled excerpts that encapsulated Geddes's philosophy, she also edited the wording of many passages, which "could only be understood with difficulty," while conserving Geddes's "picturesque style." Arthur Geddes assured her: "PG would be grateful, I'm sure, to you for pulling this off." [G.31] Lord Forrester wrote: "Despite all that I have heard you say during the work I had never realized how very far ahead of his time and our time the man was." [T.22] Percy Johnson-Marshall—later director of the Patrick Geddes Centre for Planning Studies in Edinburgh—considered *Geddes in India* to be among "the most significant books of the time." [Ek.6]

Figure 9.2   Tyrwhitt selected photographs by Anthony Denney that capture the civic beauty Geddes admired in India

Source: reproduced courtesy of Celia Denney

Linking Theory and Practice

Geddes in India was published in June 1947 just as Tyrwhitt was beginning to look for new opportunities. That month she attended an IFHTP Executive Council meeting in Paris, chaired by Pepler, which included sessions with UN and UNESCO staff to discuss possible lines of cooperation. She was thrilled when Dr. Arvid Brodersen, head of UNESCO's Social Science section "semi-seriously" offered her a job "as soon as they got the planning section straight." [6/7/47]. It had been in the spirit of UNESCO that Tyrwhitt had organized an IFHTP committee on Education for Planners at the Hastings congress. One outcome of that meeting was that educators from various countries agreed to circulate their curricula, and that APRR would act as a clearinghouse for this exchange. [B.3] At the council meeting in Paris Tyrwhitt agreed to write an article on "Training the Planner in Britain," for the IFHTP newsletter. In this piece, Tyrwhitt (1947b) shared her Geddes-inspired, pragmatic, project-based approach to planning education with an international audience. The key principle underlying her educational philosophy, she explained, was to link theory to practice: "lecture (the theory of planning), survey (the discovery of the problem), analysis (the appreciation of the problem), designing the plans (the solution of the problem), and implementing the plans (realization of the ideal) are part and parcel of one process."

In August Tyrwhitt applied Geddessian planning theory to practice in organizing an APRR conference on Survey Techniques Necessitated by the 1947 Town and Country Planning Act—which codified the "survey before plan" mantra in British planning law. Characteristically, she acted as secretary to the conference, which formed a consensus among a group of experts on the scope, procedures, and format of an "ideal county survey program" (Tyrwhitt and Waide 1949).

CIAM 6: A Turning Point

Tyrwhitt was rather absorbed with Geddes's ideas when she attended CIAM 6, hosted by the MARS group in September 1947. This event marked a turning point in Tyrwhitt's relationship with Giedion. She "subsequently became intimately involved in [his] ... life works, as translator/rewriter/editor of eight major books published in English between 1951 and 1970." [Ek.1] The first of those, *A Decade of New Architecture* (hereafter *Decade*) included a report on CIAM 6.

Rather than address any particular theme, CIAM 6 was an interim meeting, to plan for the next Congress. Members shared a commitment to revitalize CIAM—it had been ten years since the previous congress, and now "it was necessary to reformulate the goals of CIAM and to renew broken contacts" (Giedion 1967: 700)—but they disagreed over the direction of future work. One group wanted to study town planning issues, building on tenets of the Athens Charter; the other, which included many MARS members, urged new attention to aesthetics, following Giedion's call for a reintegration of architecture and the arts (Mumford 2000: 168–179).

Seven commissions were formed to discuss and report on different topics. Tyrwhitt joined Commission III, to prepare a program for CIAM 7. This commission was split into two groups reflecting these themes: Tyrwhitt served on III.A. Urbanism (planning); Giedion chaired Commission III.B. Architectural Expression (aesthetics). Tyrwhitt's contribution can be detected in the report of the Urbanism Commission, which argued that advances in physical and social planning since the Athens Charter (1933) obligated CIAM to rethink its position "in the whole domain of community planning as related to the economic and physical planning of regions and national areas" (Bridgwater 1951: 22).

MARS member Mark Hartland Thomas—also a member of the board of SPRRD—served as co-chair of the commission charged with restating CIAM's aims. It came up with a reformulation that combined both themes: planning and aesthetics. Thomas echoed in his report the holistic view Tyrwhitt had expressed in *Geddes in India*: "CIAM itself would undertake the task of imagining the new world of man as a whole, and seek to point the way towards a physical environment that would satisfy the material needs of man and give the longed-for stimulation to his spirit" (1951: 10). This wording signaled the growing influence of Tyrwhitt's synthesis of a Geddessian line of modernist planning thought—as opposed to the more formalist Corbusian line—both within the MARS group and the post-war incarnation of CIAM. Le Corbusier was a dominant force within CIAM, however, and he was commissioned to devise a "Town Planning Grid" to organize an integrated study of the aesthetics of architecture and planning at the next congress.

Giedion, UN, and UNESCO

After CIAM 6 Giedion and Fry, a member of the CIAM Council, went to Paris for meetings with Huxley, who was preparing UNESCO's budget for 1948. Anne van der Goot, a young Dutch sociologist, who was a consultant to UNESCO developing a proposal for the program on Housing and Community Planning had attended the Bridgwater congress. But Giedion and Fry urged Huxley to support a program to reform architectural education rather than van der Goot's proposal, which they feared was too ambitious. [SG.1] Giedion also took the liberty to talk with UNESCO staff about Tyrwhitt and "her" School of Planning. [SG.2] Her approach corresponded to his own philosophy of education. He confided to Tyrwhitt that he had been most interested in taking advantage of his access to Huxley to lay the groundwork for "something I thought on through years ... that we have to organize human knowledge in a completely new way." Giedion expressed those ideas—which he also referred to as "Faculty of Interrelations" [SG.11]—in the statement he drafted for CIAM to submit to the second UNESCO General Assembly: "the most vital task of this period is to learn again how to coordinate human activities, for the creation of a coherent whole. ... To achieve this, we must free ourselves from the departmentalized ... conception of education and encourage instead the understanding and comparing of the specific problems encountered and the methods evolved in the various domains of human thought and activity" (1958: 101).

Tyrwhitt gave Giedion a copy of *Geddess in India* which he read on the train from Paris back to Zurich. "I like the careful touch in the choice of passages," he wrote her. "I liked how you did it silently. Never coming to the fore. As behind a curtain." [SG.2] In return he sent her a copy of the page proofs of "Man in Equipoise," the epilogue to his soon to be published second book, *Mechanization Takes Command* (1948) (hereafter *MTC.*) "I am extremely interested what you think about these few pages, which took two years to write," he wrote in his Swiss-German inflected English, because "in a nut-shell there is condensed how I conceive the task of man." [SG.3]

Around that time Pepler nominated Tyrwhitt for a job at the UN. The position would include editing a quarterly periodical, on matters of international interest in the field of housing and town planning. Tyrwhitt asked Giedion for advice. "UNO has more money & more power as her poor sister UNESCO. But my place and maybe also yours is at the UNESCO," he replied. "UNO is big, too much involved in the politics of the day." Giedion—friends called him SG he told her—preferred to see Tyrwhitt at UNESCO. "There is creative work to do. ... And the job lying ahead of us will never be accomplished by government officials." He added: "I feel it as a kind of moral duty to help UNESCO." [SG.3,4,5]

Tyrwhitt applied for the UN job anyway. In a letter to Bauer, whom she named as a reference, she explained why:

> I am not at all certain ... that I want to work in New York. On the other hand I am quite certain that unless we can make a go of these International Organizations we are all for it again ... . My prime job, as I saw it, was to enable the Post-war world of Britain to be planned by men who had been through the war. The final Completion Course will finish at Christmas so this job will have been done then. Finally, I think it possible that we have had rather more experience, at any rate on the theoretical and legal side, than any other western country and this can possibly be put to good use abroad. [B.4]

Bauer offered Tyrwhitt strategic advice and her understanding of the roles of UN, UNESCO, and IFHTP:

> UN ... must be primarily concerned at the start with *collecting and disseminating facts* and also ... with the interchange of technicians, etc. ... [IFHTP can provide] an international meeting-ground and point of contact for all individuals and agencies responsibly concerned with immediate public policy, local as well as national ... . But the role of UNESCO seems to me quite different. Since it's not directly tied to current official policies or agencies, it is the only outfit that could deliberately set out to *broaden the horizons of the housing and planning movement*, relating it to the other elements in development of a world culture, clarifying the larger social implications of the routines decisions made by

housing and planning officials, and above all making <u>qualitative</u> judgments on experience and trends ... . [B]ut ... the UN set up would still necessarily be the core, without which neither IFHTP nor UNESCO could be very effective. [B.5]

In response to Tyrwhitt's conclusion that there would not be much "progressive construction" in Britain for the next few years, Bauer wrote: "you have no idea how the hopes of American progressives center around England these days. But if that means that you feel free to take on the international field for a while it will be wonderful for all of us." Tyrwhitt shared Bauer's letter with Giedion, who "agreed word for word." [SG.6] As it turned out, UNESCO did not approve any planning or architectural programs during the Second Assembly, and Tyrwhitt did not get the UN editing job. However, she threw her hat in the ring for future openings.

Tyrwhitt was then spending weekdays in Nottingham, where she was leading a survey team for APRR. She spent Christmas in Northern Ireland with an old friend from Dartington, the Norwegian-born weaver Gerd Hay-Edie. Gerd's husband was abroad, and Tyrwhitt played another now familiar role: caretaker for their three small children so Gerd could get some work done, starting her own textile business. Tyrwhitt enjoyed helping Gerd and being part of her young family during the holidays; Tyrwhitt otherwise had little personal life. Since Delia had been spending time in London, they had become close companions, but Delia had recently become engaged. Tyrwhitt admitted feeling "horribly sorry to lose her company—she's the perfect antidote to concentration on work." [2/28/48]

In early 1948 Tyrwhitt was stressed, conflicted, and in a quandary about her future. She consulted Giedion about studying forestry. "I liked it always, that you are close to plants & trees," Giedion replied, but he was unsure about her pursuing this specialization. [SG.7] Before making any decisions Giedion urged Tyrwhitt, who had been suffering from asthma and a hacking cough for months: "First of all you have to get your her health in order. ... You have to relax." [SG.6] Giedion suggested she spend a few weeks in the Swiss Alps. After she had some time to relax on her own, he would join her. He wanted to devote time to the "profound friendship" that was growing between them, which he saw "as the beginning of a deep mutual interest, of a common work." [SG.8] He believed in their intellectual rapport: "We see each other sharply and coldly. We have different angle to approach things. And all this does not harm a bit!" [SG.9] Fondly addressing her as "Jack of all trades," he wrote:

> I am of the same mind. You will see it soon, when you open the new book [*MTC*]. All what we need are Jacks of all trades, only I give the thing a name you probably think of as "highbrowed" – faculty of interrelationships. You have not to be a scholar ... . Its my business. You have to accomplish other things! [SG.10]

In late February Tyrwhitt flew to Switzerland. When Giedion arrived she initially felt overwhelmed in his company. He talked "of Corbusier, Picasso, James Joyce—those whom he has known intimately for many years ... the great men of our day—and yet he finds my companionship somehow worthwhile: and my criticisms of his book worthy of attention ... . But even while I still feel entirely & wholly at ease & at rest with G., I am bewildered & doubtful .... 'Can this be I?'" [2/27/48]

During this sojourn, Tyrwhitt and Giedion established a pattern of collaboration that would endure for the next two decades, while they cloaked their personal relationship behind what Giedion called an "iron curtain"—a sign also of the permeation of the cold war into the general culture. She helped him polish the English of a text he was working on; he helped her with the German version of a lecture she was to give. Together they worked on his correspondence with Mumford and her chapter on the history of town planning for a textbook based on the wartime Correspondence Course. "G was excellently critical & I feel confident now that I can adjust it." Giedion encouraged her to "write a book on tree forms & town planning & architecture—green architecture. Alas I have not the same faith in myself. ... Still, I like the idea!" [2/28/48]

Giedion brought the galley proofs of *MTC* with him and went over them with Tyrwhitt page by page. "It's a thousand pities that the translation is so bad—but the material is so good, new & interesting that I feel sure it will over-ride the cumbersome English," she thought. [2/25/48] In *MTC* Giedion broadened the theme of *Space Time and Architecture* (hereafter *STA*) (1967: vi, x) which traced the emergence of modern architecture, to show "the interrelations with other human activities and the similarity of methods that are in use today in architecture, construction, painting, city planning and science." In *MTC* he considered the even more pervasive influence of mechanization, and: "how to bridge the gap between inner and outer reality by re-establishing the dynamic equilibrium that governs [our] relationships." Giedion's use of the term "dynamic equilibrium" was inspired in part by the ideas of biologist Ludwig von Bertalanffy, a pioneer general systems theorist. [T.23] Tyrwhitt's experience and training prepared her to understand such metaphorical language; she already used the concept of ecosystems as an organizing principle in her own work. Giedion's influence strengthened systems thinking as a component of Tyrwhitt's intellectual framework.

In fact, immediately upon her return to London she supervised the work of one of her former soldier–students, John Turner, then 21, who analyzed Geddes's Notation of Life "thinking machine" diagram as "an early general systems" model. She sent Turner's draft to Giedion, who also offered his advice. [SG.12] Turner was then continuing his studies at AA and co-authored this paper with a fellow student, Keating Clay, who had been a boarder in Giedion's home while studying at ETH, and later became Giedion's son-in-law. They worked out their interpretation with the help of another student who was familiar with Chinese culture (Turner 2000: 2). Tyrwhitt and Giedion created a trans-national system of knowledge that sustained the interactive development and cross-fertilization of their ideas through their students.

Patrick Geddes: Cities in Evolution

Tyrwhitt included Turner and Clay's essay as an appendix in her abridged version of Geddes's *Cities in Evolution*—which she was then preparing—to illuminate how two young architects found "in Geddes's diagrams a scheme of thought that they could develop to co-ordinate their own twentieth century thinking."[1] She served as "general editor" of this widely read edition which she produced with Sir George Pepler on behalf of APRR, and in collaboration with Arthur Geddes, representing the Outlook Tower Association. They hoped the book would serve, as Geddes had intended, as a guide for post-war reconstruction and renewal inspired by a realistic utopianism. In this book Geddes (1949: xxx) spelled out his utopian realism:

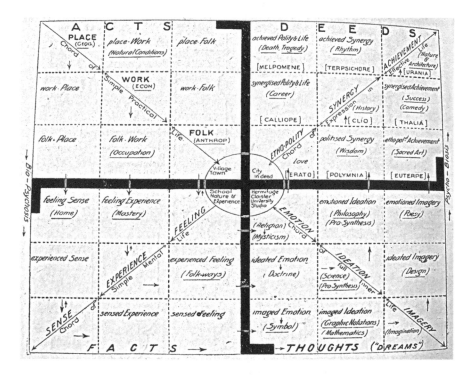

Figure 9.3   Geddes's Notation of Life diagram models a process of urban evolution

---

1   Turner included this essay in an annotated list of his publications, selected May 2007. Available at http://www.dpu-associates.net/node/100 [accessed: June 12, 2012]. Turner credits this article, his first, as leading directly to his move to Peru, where he applied Geddessian principles helping residents of squatter settlements to improve their housing (2000: 2).

> Eutopia ... lies in the city around us; and it must be planned and realized, here
> or nowhere, by us as its citizens—each a citizen of both the actual and the ideal
> city seen increasingly as one.

First printed in 1915, *Cities in Evolution* had been out of print for more than a
generation. "Perhaps it is only now ... that the time is really ripe for the reprinting
of this book," and Tyrwhitt (1949a: x) pointed in her Introduction to some of the
reasons why:

> Now that simultaneous thinking—a process that seemed almost magical when
> demonstrated by Geddes with the aid of his folded papers—has become insisted
> upon in the popular writings of every philosophical scientist. Now that sight
> from car and aeroplane, together with developments in cinematography and
> television, have made simultaneous vision a common human experience. Now
> that not only the work of the Peckham Health Centre, but almost every book
> published on popular psychology, give overwhelming evidence of the profound
> effects of the opportunities available in the immediate environment upon the
> physical and mental development of the individual.

Tyrwhitt (1949a: xi) reiterated APRR's mantra that application of Geddes's
principles to planning "can probably only be achieved by the gradual development
of a 'composite mind'... carefully cultivated by a close co-operative training
between equals at a post-graduate level." But she emphasized that Geddes "was far
more concerned that the ordinary citizen should have a vision and a comprehension
of the possibilities of his own city." To that end, he stressed "the need for a Civic
Exhibition and a permanent centre for Civic Studies in every town—an Outlook
Tower," the prototype described in *Cities in Evolution*. "This is something that,
with all our discussions on the need for and value of 'citizen participation' in town
planning has yet to be given a trial."

Tyrwhitt's edition of *Cities in Evolution* omitted five chapters that were
outdated. "These deletions are well justified," wrote Mumford,(1950: 82–83)
who considered the abridged edition "both greater and less than the original
text," "fortified" with new ingredients. "[F]or a generation that hardly knows
Geddes, except at second hand, these additions more than make up for the losses."
Mumford declared: "With the help of this supplemental fare, a representative part
of Geddes's essential thought on cities and civilizations is now for the first time at
hand." Tyrwhitt's appendix included Geddes's Notation of Life diagram as well
as the essay on it, a description of Geddes's final Dundee lecture, an excerpt from
Geddes's "Sunday Talks With My Children," and a synopsis of his life.

Significantly, she also included excerpts from a lecture Geddes gave at the
New School in 1923 that explained his concept of the Valley Section, which is
often referred to in *Cities in Evolution*, but not fully described there. Thanks to
Tyrwhitt's efforts this influential concept has since inspired many generations of
architects and planners.

Chapter 10
Trans-Atlantic Post-war Planning

New York: Spring 1948

Within a week of returning to London from Switzerland, Tyrwhitt found herself on an ocean liner to New York for an unplanned month-long visit. She was going to attend Delia's wedding; when the wedding was cancelled, Tyrwhitt made the trip anyway since her passage was paid and she thought she could help Delia. Tyrwhitt stayed with Charles Abrams (1902–1970), a prominent "houser" whom she met in London, through Bauer, just after the war. Tyrwhitt (1980: 35) recalled that Abrams "remained almost my closest friend in the USA until his death." Upon arriving in March 1948, Tyrwhitt became swept up in a whirlwind of activity with new colleagues, old friends, and the many people she met through Abrams, whose Greenwich Village townhouse was a cross-roads for a cosmopolitan crowd (Taper 1967), many among them subscribers to APRR's Information Service. At one party, Tyrwhitt was flattered to discover that architect Albert Mayer, who was then advising Indian Prime Minister Nehru on a community development project, had reviewed *Patrick Geddes in India* very favorably. [3/24/48]

Tyrwhitt stayed in constant contact with Giedion, who urged her to look up his friends; he wrote to announce her arrival to Gropius, and architect Philip Johnson, among others. Giedion also arranged for Tyrwhitt to be the first of his friends to receive a copy of *MTC*. Consequently she made it her mission to help the book find the audience it deserved, tracking reviews, lining up reviewers, and recommending revisions for future editions.

Tyrwhitt hoped to help pay her way in the US by giving lectures. Thanks to Abrams, who taught at the New School for Social Research, she was hired to give three lectures there: Surveys for Planning, Central Area Redevelopment, and Neighborhoods. Bauer attended Tyrwhitt's first lecture. The following day she took Tyrwhitt on a tour of housing projects, and their "conversation ranged over the wide and international field of housing." Tyrwhitt observed: "'houses' versus flats' is as perennial and as unsatisfactory an argument with us as 'public versus private housing' is here." That led to "an interesting discussion on the possibilities and values of accommodation shared in common between groups of neighbors as the grass roots of democratic training in citizenship." Tyrwhitt added: "Catherine and I were all for a predominance of high density row building with very small private open space—and 'outdoor room' to each house—and ample public open space near by." [T.24]

Abrams featured Tyrwhitt's observations in his column in the *New York Post*. "Miss Tyrwhitt seemed particularly disturbed about the lack of the neighborhood

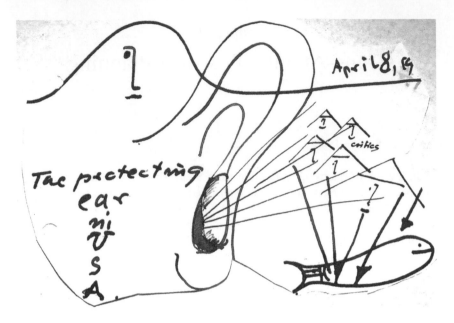

Figure 10.1   In 1948 Giedion relied on Tyrwhitt to be his "eyes and ears" in the United States

atmosphere in the projects," he wrote. She also noted the danger of "regimentation" in large projects. Europeans had learned in the war that constant supervision leaves no room for children's play. Tyrwhitt explained: "The only playground that could really keep the children off the streets was the bombed site, where there was adventure accompanied by dirt and danger." Noting that the only play area in Stuyvesant Town—where 8,800 apartments for returning veterans were being built on 75 acres in blocks of 12 and 13 stories—was "a supervised fenced-off asphalt playground with a formal array of equipment," she predicted: "I wouldn't blame [the children] a bit if they sought refuge in the streets nearby." "More frank criticisms like these are needed, more interchange of ideas before our mistakes become encased in brick and mortar," Abrams wrote. [T.25]

Abrams also introduced Tyrwhitt to New School president Bryn Hovde to discuss the possibility of Tyrwhitt returning for six months to set up a planning program there based on the School of Planning diploma course. She reflected:

> I must say I'm definitely attracted to the idea of getting a course started here. One feels that if some ideas were put into people's heads here something really could happen—they are prepared to act. At home I know we have to mark time for the next few years & maybe it really would be worth trying to create something here. [3/31/48]

By the end of her first week in New York, Tyrwhitt had received invitations to speak at: Yale, Harvard, MIT and Columbia; and to visit Mumford. Exalted, she exclaimed: "A bit more solid work with myself & there will be something to give out but I know I am onto something sensible." [3/26/48]

New England

The social swirl continued in Boston. Tyrwhitt especially enjoyed having lunch with Gropius and his wife Ise at their home. "We had so many things to discuss— German planners, CIAM summer school, Giedion's book, a new series he wants to get out—& America in general & planning & architectural training & practice in particular." [4/10/48]

One outcome of her visit was an invitation to contribute to the first post-war issue of the student journal *TASK 7/8* (1948). The co-editors of *TASK 7/8* were Martin Meyerson, then a graduate student in planning at GSD, and his wife, Marjorie, both disciples of Bauer. The Meyersons dedicated this issue to reconstruction. Bauer (1948: 6) noted in her introduction that transnational organizations such as IFHTP, UN and UNESCO were instigating the post-war "revival and intensification" of "broad–based international fellowship and cross-fertilization in this field."

In her article, "Reconstruction: Great Britain," Tyrwhitt (1948a: 20) voiced the frustration she had shared privately with Bauer: "Here at long last, we have within our grasp the means to plan, in the shape of the 1947 Town and Country Planning Act, and it turns out to be but a mirage—an image of what can be—one day, not now, not for a long time yet." On the other hand, there was a silver lining: "perhaps it is as well that economic conditions will entirely prevent the Abercrombie stamp being given concrete expression in every other town up and down the country." In this regard she concurred with Mumford (1946) who also warned against the particular danger posed by wholesale application of stereotypical solutions such as the garden city model based on the neighborhood unit.) She concluded hopefully: "after ... patiently cultivating our garden we may again have something worthwhile to put forward on imaginatively constructive planning."

From Boston, Tyrwhitt went to visit Mumford in New Hampshire. She consulted him about her proposed planning course at the New School; it meant a lot to her to receive his support. Mumford also gave his blessing to her idea for transcribing, for publication as a book, Geddes's proposal for a university for Central India, at Indore. It is probably during this visit that Mumford made a donation to APRR in memory of his son, who had been killed; he authorized Tyrwhitt to make it a revolving fund for needy students.

CIAM International Summer School

Concomitantly, Tyrwhitt also worked on both sides of the Atlantic as a liaison between Giedion and the MARS group to help plan CIAM's first educational undertaking, an international summer school to be held in London that summer. The idea was to build on a program that the AA school had inaugurated. When that fell through Tyrwhitt's relationship with Giedion provided the glue that kept the project alive. At Tyrwhitt's instigation, Giedion asked Max Fry to direct the program. "Fry, you and at the outskirts myself ... It may be an experiment worth trying," Giedion wrote. "We will develop ... a very good program for educational reform in architecture." [SG.13]

Giedion was determined that CIAM show leadership with this experiment. "If we are successful UNESCO will follow," he thought. [SG.14] UNESCO staff had advised Giedion to work with IFHTP and UIA to revive the proposal that he and Fry had presented following the Bridgwater congress. Now Giedion enlisted Tyrwhitt as an ally in his negotiations with IFHTP. He saw her role as: "active, but not imitating men's logic & rationalistic behaviour. But being a friendly, smooth female, behind the man's activity." [SG.11] Tyrwhitt accepted this role in order to help Giedion, when IFHTP convened in Zurich in June.

She then applied her prodigious energy and efficiency as assistant to Fry in directing the first CIAM summer school, which ran for six weeks beginning mid-July 1948 (APRR 1948a). In focusing on "architectural aspects of the central urban replanning" this summer school served as a precursor to the urban design program that Tyrwhitt assisted CIAM president Jose Luis Sert and Giedion to establish at Harvard a decade later.

New York: Fall and Winter 1948–49

Fortunately the seeds Tyrwhitt had planted in New York bore fruit: she was invited to return to the New School to lecture—on "Town and Country Planning in Britain and the US"—during the autumn semester, and to work with Abrams developing the proposed planning course. This opportunity propelled Tyrwhitt into a new role, as a conduit for the transnational flow of planning ideas, at a time when such exchange was hampered by the high cost of travel, and restraints on the mobility of people and capital.

Tyrwhitt gave her first lecture on the concept of regionalism—"mostly Prof. Taylor with a little Fawcett & Dickinson and a flavor of Geddes." [10/8/48] Those lectures, which were held in the evening, were "quite something," Tyrwhitt (1980: 36) recalled: "the regular front bench of my audience included Lewis Mumford, Charlie, Albert Mayer, Charles Ascher [then adviser to Huxley on UNESCO programs] and Clarence Stein were also frequent visitors. ... Once the class was over Charlie would invite the group to [his house two blocks away] and we would have drinks and nuts and talk away in his enormous drawing room until ... late."

Tyrwhitt was honored to be invited to the home of architect Clarence Stein (1882–1975), a co-founder of the Regional Planning Association of America (RPAA), who convened a meeting to mark the 50th anniversary of the publication of Ebenezer Howard's *Tomorrow: A Peaceful Path to Real Reform* (1898), and to discuss reforming RPAA. The distinguished gathering included both leaders of the "old guard" and representatives of the "young planners," who were in town to attend the annual American Society of Planning Officials (ASPO) conference, which opened there the following day. [10/10/48] Just as Mumford had introduced Geddes to RPAA members in 1923, Tyrwhitt introduced Geddes's ideas into the discussion at the ASPO conference session on Training the Planner, during which "the growing pains and adolescent confusion of the new and rapidly growing profession" were evident (Taylor 1949: 328). The conference report singled out Tyrwhitt's remark, that "training for town and country planning in [Britain] even many years before the present Labor government, had not been geared to 'what exists' but to 'what might be.'" Rehearsing a theme from her introduction to her edition of *Cities in Evolution*, Tyrwhitt also asked: "When does citizen participation come in?" Here, she referred to Geddes's proposals for a "civic museum" and "outlook tower" as "worth more thoughtful exploration in meeting today's needs and wants" noting, "I tried to point out the value of the simple ideas that could be readily visualized by the public." A student at Columbia University who had corresponded with Tyrwhitt during the war years about APRR as a model to replicate in the US, recalled that her views "found ready echo among the 'young planners' searching for ... direction." [E.7]

Regarding the proposed New School planning program, Tyrwhitt worked with Abrams, Mayer, and Mumford. They considered two options: a comprehensive graduate course, and an annual lecture series with guest speakers from Europe. Abrams and Mayer were in favor of continuing to invite guest lecturers "to make their headquarters at the New School and permeate a wider public." They were impressed by the invitations Tyrwhitt received: "a week in Chicago; to open a conference in Boston and ...2 Harvard seminars." Tyrwhitt thought: "They ... think that this opportunity of spreading a fresh point of view may be more important at the moment ... than an enclosed school." While this would preclude a job for her, she claimed to be indifferent. "In a way I do not want again to get tied to a job ... . just now I revel in a sense of (limited) freedom." However, she argued in favor of the full course, "mainly as an experiment in integrated education." Mumford favored the guest lecturers, but suggested a third option: a seminar using all the resources of New York City. [11/2/48] Tyrwhitt's final proposal included a fourth idea: creating a Social and Physical Planning Forum that would meet regularly "to discuss important planning projects that are now being handled in various parts of the country." Based on those discussions, the Forum would make recommendations for a US planning policy. Without such a

policy, "it's useless to drill up an elaborate training course here," she concluded.[1]
[1/17/49] [T.14]

Lecture Tours to Chicago and Cambridge

Tyrwhitt made her out of town trips in between her weekly lectures. Her hosts in Chicago were ASPO director Walter Blucher and Herbert Emmerich, director of the Public Administration Clearing House (PACH); they arranged a full schedule of meetings, lectures, and field trips. Rexford Tugwell, the New Deal "brain-truster" who had recently started a research-based planning program at the University of Chicago—where Meyerson now taught—arranged meetings with his staff, and a lecture to the Social Science faculty. (She spoke on "Postwar trends in England in Planning and New Town Development.") She also visited Reginald Isaacs (1911–1986), whom she became friends with in 1945 when they had "agreed on a critical attitude to the 'neighborhood unit;'"[2] he was now director of planning at the Michael Reese Hospital. Tyrwhitt liked Isaacs' modernist master plan for the hospital, on which his mentor Gropius had consulted. "I'm sick of suburbia and plans for suburban new towns," she declared. Chicago may be "about the most outrageous urban mess in the world … but it has courage & does try out things." [11/9/48]

Tyrwhitt gave an informal talk at the Institute of Design, where several of Giedion's friends were on the faculty, and was fascinated by a special course given by "the Dymaxion man," Buckminster Fuller: "I got a private lesson with copious illustrations of the behaviours of four great circles & the tetrahedron & the octahedron & their relationship to their bastard cousin the cube & how the insistence on the cube has twisted our outlook ever since Euclid … . Of course it all rang absolutely true as one comes more & more to hexagonal & triangular solutions of all problems in planning … & also it fits in with botany." [11/10/48]

Tyrwhitt spent hours fiddling with the set of geometric parts that Fuller gave her. Assembling them helped her to better understand his ideas, and she was thinking about geodesics when she spoke on "New Patterns for Greater London" at the National Municipal League conference in Boston (see Tyrwhitt 1948b). During a lull she began doodling and devised a hexagonal design for a decentralized metropolis "as an aesthetic exercise." [11/24/48] She was feeling creative, she was happy, she was living with and surrounded by friends in New York and elsewhere, and after leading seminars at Harvard and MIT, she felt stimulated and welcomed by the academic planning community.

_____

1 The outcome of these proposals is not clear.
2 Tyrwhiit's critique of the neighborhood unit is discussed later.

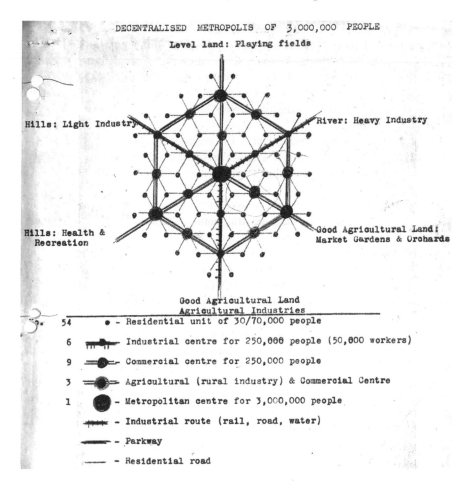

Figure 10.2   Buckminster Fuller inspired Tyrwhitt to develop a model for a de-centralized metropolis based on a hexagonal grid

Translating Giedion

"I would like to work next fall in USA if you are there," Giedion wrote. "I would like to help to reform MIT. ... But you will see the situation and give me your opinion." [SG.14] Tyrwhitt tactfully explored prospects for Giedion at MIT over lunch with William Wurster, then dean of the School of Architecture and Planning, and John Burchard, then dean of the School of Humanities. Wurster arranged for Tyrwhitt to tour the new MIT dormitory designed by Alvar Aalto, which was then under construction. She was thrilled by the building, and wrote enthusiastically about it to Giedion. [11/26/48] "What you said was my immediate feeling when I saw first the plans ... in 1947," he replied. [SG.15] Giedion proceeded to quote her description—"well protected by an iron curtain"—in the new chapter he wrote on Aalto for the eighth (1949) edition of *Space Time and Architecture (STA)*.[3]

The line distinguishing her words from his was blurring. Earlier, Giedion hinted that he could use Tyrwhitt's help with this edition of *STA*. Now he was behind schedule. "It goes so slowly with a man to whom you have to suggest the English words," he wrote to Tyrwhitt in New York:

> I hesitated but now I am sending you ... two fragments ... . Will you go through
> them and may I send the next? ... Writing must reflect the tone of the voice ... But
> to work in a foreign material is still extremely dilettante;. ... Be nice, Jaqui." [SG.15]

Tyrwhitt was flattered that Giedion used her words and gladly took on the translation work for him. He sent her fragments of text, as he completed them: "You know the sound of my voice and you can certainly rearrange it in the right manner." [SG.16] He was delighted with how she "smoothened the translation," which she left colored with "Giedionese." Giedion asked her advice: "Is it possible to call the whole chapter 'ALVAR AALTO, the contemporary and the primitive' and could we not write for *your* subtitle: "The complementarity of the differential and the primitive?'" The word "complementarity" referred to "the great invention of the physicist Bohr." But Giedion was unsure about "differential." [SG.17, 18] (Tyrwhitt re–titled the chapter: "Alvar Aalto: Irrationality and Standardization" and the section: "The Complementarity of the Differentiated and the Primitive.")

The significance of the 1949 edition of *STA* was that Giedion's new chapter on Aalto offered an "organic" aesthetic and integrative approach to planning and design as "a powerful spur to thinking about an alternative to function" at a time when modernist architects and planners were searching for a "new humanism" (Bullock 2002: 48). Giedion could not have presented this timely prescription without Tyrwhitt's ability to clarify his often vague and formless texts, and her willingness to do so without credit or—in this case—compensation.

While working on the Aalto chapter, Tyrwhitt received a draft article from Giedion on Moholy-Nagy, and assumed he meant for her to "adjust" this text, too. Giedion later claimed this was a misunderstanding. "I felt very guilty, when you rewrote it," Giedion wrote, "but please believe me, I do not regard you as a translator house." [SG.19] He also asked Tyrwhitt, while she was still in Cambridge, to represent him to Harvard University Press (HUP) to ensure the layout of *STA* was as he wished. HUP found it helpful to work with Tyrwhitt, and wanted to continue. By then Tyrwhitt was about to return to England, but she did continue, translating Giedion's new material—significant additions in each case—for the third (1953), fourth (1961), and fifth (1966) editions of *STA,* and working with HUP to prepare each for publication. Giedion did not acknowledge her contributions until the fifth, final edition (1966: v).

---

3   Giedion also published this text as "The Undulating Wall," *Architectural Review* (Feb 1950): 77–84.

Tyrwhitt was happy, at least initially, to work anonymously for Giedion because she believed in him and in the relationship she imagined that they were building. In order to further their collaboration, Tyrwhitt suggested: "I think somehow CIAM must have an assistant secretary. Surely that's the easiest way." [SG.20] Giedion agreed. Knowing that Tyrwhitt was going to spend a weekend with Gropius in a few weeks, Giedion wrote in advance, in his capacity as CIAM secretary, to suggest that she represent Gropius, who was not well, at the next CIAM Council meeting, to be held in Paris in March.

She worried, though, about her own professional future. Given this uncertainty, Tyrwhitt made the most of her time in New York. She assiduously nurtured her personal connections and in doing so tried to help other people, too, for example: promoting Rowse's schemes; and arranging a meeting between Forrester and the Rockefeller Foundation. She submitted her own—ultimately unsuccessful—application to Rockefeller "for a grant to do the Geddes book on University Renewal."[SG.21]

Tyrwhitt also continued to study, especially about modern art. At Meyerson's suggestion, she sat in on art historian Meyer Schapiro's "dazzling" lectures on modern painting at the New School, and she spent hours at the Museum of Modern Art, art galleries, and at the homes of friends with private collections. Architect Knud Lönberg-Holm (1895–1972), a member of the New York CIAM chapter, told Tyrwhitt about Norbert Wiener's work on "teleological mechanisms" at MIT—the type of feedback essential for living things to achieve "dynamic equilibrium." This inspired Tyrwhitt to further investigate Wiener's ideas, which resonated with what she was learning about modern painting. These informal studies, in turn, deepened her appreciation for Giedion's ideas about the mutual influences of science and art—a central theme in *STA*.

In late January 1949 Tyrwhitt returned to the Gropius' home with the Wursters. Over tea they "talked of architecture and personalities and the future of various countries and the planning and housing movements in the US … . I kept as it were sitting back and watching myself there, in that room, with those people: knowing that I seemed to fit in quite well, wondering why and how it had happened." [2/10/49] Clearly, her relationship with Giedion had a lot to do with it. But Tyrwhitt, *from behind the iron curtain,* was joining this elite circle as an individual, and not as part of a couple. "The Indians believe that a woman between 40 and 60 can shoulder more than a man at any time," Ise Gropius later told her. "So remember that you are coming into your very best years now." [SG.22] Words which proved to be true, although Tyrwhitt remained conflicted about her own ambition.

Tyrwhitt's last lecture at the New School, the reprise of one given by Geddes on his visit there in 1923, paralleled her work for Giedion in an interesting way: it involved a translation, not merely a transcription of Geddes's words. Tyrwhitt rewrote Geddes's lecture based on some rough shorthand notes she had found: "It was some job though as the shorthand writer knew nothing of planning and took down things phonetically as Geddes mumbled them swiftly into his beard." Tyrwhitt interpreted Geddes's words much as she clarified the meaning of Giedion's.

The Town and Country Planning Textbook

When Tyrwhitt returned to London she completed her last project for APRR, which closed in 1950: editing the *Town and Country Planning Textbook* (1950). She reckoned: "It's reasonably appropriate to bill the Textbook as APRR's swan-song. On the whole it does contain the raison d'être of our existence, and the proof that it was worth it." [T.26] Tyrwhitt stated in her preface that the "remarkable success" of the Correspondence Course encouraged its organizers "that the general publication of such a course of study would meet growing demand" (1950a: xv). Preparing this textbook—the first of its kind in Britain—involved revising and supplementing the original lectures to reflect the new requirements created by passage of the 1947 Town and Country Act. New material from several social science disciplines, in particular, was added. Tyrwhitt explained why APRR was credited as the editor: "Just as Planning is not the work of one brain but rather the result of a joint effort of many individuals trained previously in different specialist fields, so the evolution of this book should be recognized as the product of such a team." Tyrwhitt deserves credit as the leader of this team effort that produced a collection that represents the "sum of town planning theory and practice" at that time (White 1974: 45).

Tyrwhitt was explicit about the synthesis of Geddessian and modernist social-aesthetic ideals this collection represented: "The Patrick Geddes's triad 'place, folk, work' and the four points of the CIAM *Charte d'Athenes* 'living, working, developing mind and body, circulating' are fully treated, and, though the purpose of the book is to impart technical information, there is a constant warm under current of enthusiasm for the well-being of a lively and diversified humanity" (APRR 1950: 1). Her contributions include "Chapter 6, Society and Environment: A Historical Review;" "Chapter 7, Surveys for Planning," and the Bibliography. Those topics represent three facets of Tyrwhitt's scientific humanist conception of planning as:

- grounded in an evolutionary macro-historical theoretical perspective;
- based on empirical research, using the survey method both as an analytic tool and as a means of civic engagement in the planning process; and
- a holistic, integrative process, that requires as a corollary the coordination and classification of different branches of knowledge.

Tyrwhitt's contributions conveyed the potential of an ecological approach to community design that integrated economic, social, and land use issues with aesthetic and spiritual/emotional concerns in the context of regional planning.

In Chapter 6, Tyrwhitt (1950b: 96) conceived of the history of town planning in broad ecological terms as: "the study of man in relation to his environment." The study of town planning should begin, therefore, with analysis of "the universal needs of community life." The question is: under what conditions do communities evolve to higher forms of civilization (organized in cities)? In tracing the evolution of civilization/cities from their ancient origins Tyrwhitt paid particular attention to the urban core—the wellspring of civic life—and then focused on urbanization in England.

Her discussion of Modern Trends focused on the evolution of the Garden City concept. She criticized the conservative tendencies in the Garden City movement and called for a more creative approach to civic design, grounded in love for *existing* places: "Each place has a true personality; … which it is the task of the planner, as master-artist, to awaken" (1950b: 139). Tyrwhitt argued that "planning exponents have tended to divide into two classes," and only one is heir to the Geddessian tradition:

> The first link *Folk* and *Work*. They believe that the best life can be lived in a new town of limited size closely related to sufficient industry to provide its population with their daily bread. … The second link *Folk, Work* and *Place*. They are convinced of the inter-relation of history and environment with man's daily life, and that the problems of congested, unhealthy, over-grown cities can only be solved when these cities are considered as a whole, in their regional setting. (1950b: 139)

Tyrwhitt gave the final word to Giedion (from *STA*)—thereby establishing him as an heir to the Geddessian tradition:

> The modern town planner is not primarily concerned with architecture. He seeks to discover how the town came into being and how it has reached its present growth. He wants to know as much as he can of the site and of its relations to the surrounding district and the country as a whole. Above all he studies the different categories of people who haave to be accommodated, each according to their manner of life. (in Tyrwhitt 1950b: 145)

Chapter 7, which Tyrwhitt compiled, provided detailed guidelines for undertaking planning surveys, and provided the first textbook discussion in English of methods for using transparent thematic map overlays (Collins, Steiner, and Rushman 2001). While not unique, Tyrwhitt's overlay method was widely adopted in Britain and became a standard feature of ecological planning methods in North America, including those popularized by Ian McHarg in the 1960s. It represented a predecessor of automated geographic information systems (GIS), which use digital map layers rather than paper (Chrisman 2005; Cloud 2002; Steinitz et al. 1976; Malczewski 2001). Tyrwhitt (1950c: 146–147) stressed, moreover, that the survey method—in addition to serving as a means for collecting, coordinating and visualizing information—could make "community planning a truly democratic process … .The community must see that its varied problems are inter-related and it must study many of them concurrently."

APRR stated on the cover that the textbook's "greatest value is probably as a reference book both for the practicing planner and for other professional men in allied fields." The Bibliography was organized according to APRR's unique library classification system, which was designed explicitly to facilitate integrative study of human settlements as complex systems by specialists as well as lay community members.

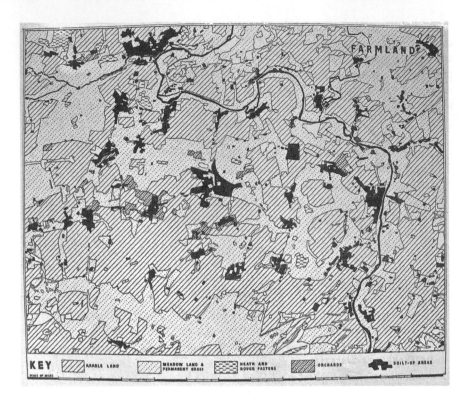

Figure 10.3   Tyrwhitt's method for using transparent thematic map overlays was widely adopted in Britain and became a standard feature of ecological planning methods

Source: Originally appeared APRR (1950 p. 167)

Part IV
1949–1956

Chapter 11

Planning for Real and Ideal Cities

Size and Spacing of Urban Communities

When Tyrwhitt returned to Britain in February 1949 she completed work on the Nottingham survey she had organized and supervised for APRR. The work, which was mainly done by local authorities and universities, involved the collection and presentation of a complete town planning survey (population 500,000), and included a series of large maps, 112 volumes of reports, and detailed analysis of land use using tabulating machines and mechanical punch cards. [T.27] One thing she hoped the survey would make clear was "HOW MUCH and HOW MANY of this and that kind of buildings (incorporating services) a total population of about half a million (including hinterland) really does support." [B.6]

Tyrwhitt was interested in using the results of such surveys to refine her critique of the garden city (of about 50,000 people) divided in neighborhood units (of about 5–10,000 people), which was then being promoted as an ideal standard for post-war development. She had conceived of an alternative model for metropolitan planning based on the hexagonal geometry that Buckminster Fuller had inspired her to think about. She developed these ideas in the context of a lecture she gave, "The Size and Spacing of Urban Communities," at Vassar College, in Poughkeepsie New York in February 1949. [T.28] Tyrwhitt circulated the paper based on this lecture to a small group of influential friends, including Abrams, Bauer, van der Groot (who was then director of the Housing and Town and Country Planning section of the UN Department of Social Affairs), and Mumford. This led to its publication in the *Journal of the American Institute of Planners* (JAIP) in the summer of 1949. The APRR Information Bulletin concurrently published a synopsis of the ideas, along with Tyrwhitt's hexagonal diagram. Her argument followed two lines of reasoning, about people's needs, and current trends affecting development.[1]

Adopting American-style language, Tyrwhitt suggested that the planner's aim to provide "equal opportunity for full individual development" was comparable to other democratic efforts to provide equal opportunity for education, health, and housing. The US Housing Act of 1949—for which both Bauer and Abrams lobbied— pledged a decent home for every American family. While the garden city type of community catered well to the needs of many families with children, Tyrwhitt argued that it could not satisfy the "twentieth century needs of a family as individuals, throughout their life." Furthermore, rather than impose a concept of a perfect city "upon a reluctant world," she thought, the planner must understand

---

1   The following summary is based on APRR (1949a: 1) and Tyrwhitt (1949b).

current trends and "recognize the dynamic within each that can, by *wise and sympathetic guidance*, make the lives of the inhabitants of a given area richer and freer" (emphasis added). Current trends pointed toward continued decentralization, reinforcing urban decline. Yet "our civilization ... grew from within the great cities of the past. How can we interpret these trends to develop the civilization of the future?"

She proposed replacing the Garden City ideal with "a sounder one" based on realistic premises. Since the trend toward segregated suburbs was "to some extent natural," and "one cannot change human nature overnight," a physical planner could not do much to solve that problem. That said, Tyrwhitt proposed a "mixed neighborhood" with a population of 15,000 as the basic "social unit," which, when paired forms an "urban unit" of 30,000, organized around a high school, where "true social integration will become increasingly easy and normal." She proposed using the "social unit" and "urban unit" in place of the "'neighborhood unit' with its ... almost universal, connotation of a segregated community."

Tyrwhitt based her ideal metropolis of a million people—"a descendent of the satellite town, the linear city, la ville radieuse, and other theoretic planning patterns"—on a hexagonal grid containing 30 urban units, 24 of which were equidistant from both a large, diverse industrial area and a lively commercial center. These urban units were connected via highway to a central city that featured a business district, university, cultural facilities, reference libraries, specialized hospitals, and the most advanced products "of our civilization." Here residents from all parts of the metropolis could "meet easily and freely" to participate in the development of the future civilization.

Tyrwhitt's effort to re-interpret the Garden City and Neighborhood Unit in modern terms constituted a substantive contribution to the tradition of geometric planning concepts generally, and to hexagonal concepts in particular. Tyrwhitt's contribution to this line of thought is two-fold: a pragmatic way to redirect decentralizing trends away from producing sprawling, socially and spatially segregated suburbs, and toward the development of compact, diverse communities via the "urban unit;" and to emphasize the nodal role of the large city as the *cultural* center of the metropolitan region.

Tyrwhitt continued to develop these ideas, but not right away. However, they did immediately enter transnational discourse via Tyrwhitt's correspondence with Mumford and Bauer, which she shared with Giedion, who in turn had her paper translated into German for publication in a new Swiss journal.[2]

---

2 Tracy Augur—an RPAA member and proponent of decentralization—wrote a response to Tyrwhitt's article, which he felt incorrectly inferred that Howard's Garden City idea did not admit a role for larger urban centers. He acknowledged: "Miss Tyrwhitt has written in the spirit of the Garden City pioneers and has performed a service by again calling our attention to the need of city forms that will fill the many and varied needs of an urban society" (1949:42–43).

Irons in the Fire

Tyrwhitt was unsure of her next move, which depended on whether Giedion would be invited to MIT in the coming year. "Future plans don't exist yet," she informed Bauer:

> I'd like to come over again next winter if I can. ... On the other hand I'm going to meet some chaps this week-end who want me to team but with them here— but I don't really think I'm ripe for that just now. We'll see—its rather fun to feel free! [B.7]

In April, one possibility of returning to the US materialized when she received an enquiry from Vassar about her interest in serving as a consultant to advise the college on the development of a program of community studies that would utilize a college-owned estate. Tyrwhitt came up with the idea of equipping Vassar's facility with the latest "tabulating machinery" and a staff of statisticians who would scour the existing body of planning surveys for useful data and prepare a series of comparative analyses. This work would focus initially on the US, but ultimately would become international. Tyrwhitt explained to Bauer, whose advice she sought: "I feel very strongly that we spend too much time rushing after new material and have failed to perceive the lessons of what we have already collected." [B.8]

Bauer, a Vassar alumna, had also been asked how to use the estate. She supported the type of center Tyrwhitt proposed, but had doubts about setting it up in such a remote location. Bauer suggested that Tyrwhitt consult with Meyerson who was interested in that kind of research operation, and recommended that Vassar tie into the on-going effort of a consortium of planning schools to secure Ford Foundation funding for urban planning research. [B.9] Tyrwhitt followed that advice.

Tyrwhitt's proposal to automate an urban information system was in the forefront of this trend. She was aware of a similar project to form an international council for building documentation then underway under the auspices of a committee of the UN Economic Commission for Europe (UNECE). Yugoslavian architect Ernest Weissmann (1891–1969), Director of the UNECE Industry and Materials Division, based in Geneva—and a CIAM member—encouraged CIAM to become involved. Tyrwhitt knew that Wells Coates (1895–1958), who chaired the CIAM Commission on Industrialization of Building—one of the seven created at Bridgwater—wanted to establish a relationship to the UNECE initiative. [T.29] Tyrwhitt had pragmatically formed a partnership with Coates, a co-founder of the MARS group, during this period. Tyrwhitt's association with Coates solidified her already strong ties to the MARS group. She was then helping them plan the second CIAM summer school, which she would again assist in directing.

In early July Tyrwhitt heard from Giedion that he would return to MIT as a Visiting Professor for the 1950 spring term—the result of efforts he had made to be in the US if she were there. "You must tell me soon," Tyrwhitt wrote Giedion, "what I must do myself." [SG.23] She was then back in Ireland helping Hay-Edie. Buoyed by the MIT invitation, Giedion felt more optimistic about CIAM 7—he had been annoyed by the continuing lack of cooperation and activity among the members— and encouraged Tyrwhitt to attend the congress, which was to be held in Bergamo, Italy in late July. He even bought her new clothes so she would look her best (she tended to not fuss about her appearance). Giedion also invited her to join him in September in Les Ezyies, France, where he would study the Paleolithic cave paintings for his new book on the ancient origins of art and architecture.

Tyrwhitt kept an eye out for articles about Paleolithic artworks, especially those unearthed around Les Ezyies. The Lascaux cave complex, discovered in 1940 and only opened to the public in 1948, had sparked interest in those stunning paintings. She soon discovered personal connections to the conversation about this topic; Bunty Wells, for example, had studied the Lascaux cave paintings and gave Tyrwhitt her notes and contacts. Tyrwhitt wanted to share this information with Giedion, and to learn enough to be able to help him "in some selection of words" for his MIT lectures. [SG.23]

CIAM 7—Town Planning Grid

Tyrwhitt brought the material she had collected on cave painting to Bergamo for CIAM 7. This was the first Congress she attended in her new role, as acting secretary to the CIAM Council. She had assumed this role at the Council meeting in Paris the previous March, where Tyrwhitt had represented Gropius—as she had engineered with Giedion. At that meeting she had volunteered to take on "an enormous amount of work" to help Giedion, who could no longer do the job of General Secretary alone. In gratitude Le Corbusier had bestowed on Tyrwhitt the soubriquet "Valentine de Boston." [T.30]

At CIAM 7 Tyrwhitt helped CIAM, collectively, "find the right words." One of the main topics of discussion at this congress was the "Town Planning Grid" (*grille CIAM d'urbanisme*) that Le Corbusier and his fellow members of the ASCORAL group had developed as the presentation format for projects submitted by each member group. Each Commission considered the value of this Grid while discussing an agenda for the next congress. Le Corbusier headed the Urbanism (planning) Commission that Tyrwhitt served on. They accepted use of the Grid and also dealt with Le Corbusier's proposal that CIAM develop *L'Charte de L'Habitat* based on the Grid to complement the Athens Charter. Tyrwhitt identified a problem with this terminology:

> This can be translated so variously that it has been left as it stood. It should not be transformed [literally] into a "Charter for the Habitation." The MARS group has accepted a wide interpretation of the words and is working on "civic centers." (in Gold 1997: 209)

Figure 11.1 Le Corbusier gave Tyrwhitt this sketch. The inscription reads: *Pour Valentine de Boston. La perle de vertus cardinales. Avec tout mon amitie. Lendemain de Hoddesdon, Juillet 1951*

As early as July 1949, Tyrwhitt attempted to mediate potential conflict arising from the difficulty of finding an English equivalent of *L'Charte de L'Habitat* by forging a conceptual linkage with the term "civic center." These emerged as the competing themes for CIAM 8, which the MARS group proposed hosting in England during the summer of 1951, to coincide with the Festival of Britain. [T.31]

Tyrwhitt also offered to translate and edit a short version of the proceedings of CIAM 7. Giedion looked forward to their collaborating: "The first thing on which we will work together will be the pamphlet of the CIAM Congress. I would like to write the introduction & select with you the papers so that it will be something readable." [SG.24] He instructed Tyrwhitt to negotiate with Lund Humphries to publish the book per his requirements. She relished the assignment; it not only kept her engaged with Giedion, but also required correspondence with Sert on CIAM affairs. Sert believed the continued existence and evolution of CIAM as a worldwide organization depended on demonstrating the relevance of the organization through publications. This had proved to be easier said than done. Giedion had been commissioned at Bridgwater to produce an account of work by

CIAM members during the decade since the previous Congress in 1937; he was still struggling to assemble enough material for a book.

After Bergamo Tyrwhitt was exhausted, but she threw herself into running the second CIAM Summer School with Fry. This course continued a discussion that began at Bergamo that explored how town planning could provide the context for collaboration between artists and architects. The term concluded with a symposium on Painting, Sculpture ,and the Architect at RIBA, attended by over 600 people. Tyrwhitt's role was to sum up, and "made a success of it!" To act as organizer and synthesizer was her forte. Coates, however, wanted to get Tyrwhitt focused on their partnership, "doing real jobs and not running around organizing things." Tyrwhitt admitted: "in a way he's right;" but she was waiting to get the cue for her next move from Giedion. [SG.25]

In September, after two weeks with Giedion in Les Ezyies, with no visible means of support and no home of her own, Tyrwhitt returned to Ireland, where she received room and board in exchange for childcare. This time, Hay-Edie went away for a few weeks, and Tyrwhitt felt overwhelmed by having to care for four small children by herself. Her asthma flared up, and her mood turned gloomy. "I have many things I should do and no physical powers to do them," she wrote Giedion. "I suppose really the best thing would be to do NOTHING for these 3 months [until she could join him in the US] —except housework and the children … It may be very wrong of me to say this—but I think this background of

Figure 11.2   Tyrwhitt (center) and Giedion (left), ca. 1949

uncertainty and 'wishful thinking' is part of my trouble." [SG.26] Giedion replied: "Go away on a place where people take the necessary care of you," and offered to pay if money was a problem. "Maybe your present state can also be explained because ... you are not accustomed ... not to be busy, you did not relax for a long time." [SG.27] His letter touched Tyrwhitt. "It is so strange ... to read kind and friendly and helpful words about <u>me</u>, ... You have seen (and laughed at) the way that I am far more accustomed to have to deal with and look after people who are even more foolish than myself." [SG.28]

Critique of the CIAM Grid

Energized by Giedion's expression of concern, and with the burden of daily chores lifted by Hay-Edie's return, Tyrwhitt worked on the proceedings of the Bergamo Congress. She also wrote a brief report on the Congress for the APRR Information Bulletin, in which she critiqued both the CIAM Grid as a presentation tool for town planning schemes, and the plans that were presented at Bergamo. While "there is almost universal agreement that some system of standardization is advantageous," she argued, the CIAM grid, "with its architectural emphasis on building construction—was far less successful for town plans, as [the British] understand them," than for smaller scale civic design projects—"which most of us would consider the sole province of the architect." This distinction between civic design and planning "now familiar ... in Britain — has scarcely been posed elsewhere," she observed. Most British professionals accepted that planning was a continuous process with "recurrent interpretations of survey material" more or less replacing the now discredited "static 'master plan.'" This situation became possible once such a planning system had become an accepted role of government. Now that the "flexible plan" is part of the "normal environment," the architect's role is to give that plan "plastic reality." Tyrwhitt disparaged, however, the "lacunae between elementary statements of physical survey and the plans and elevations of civic buildings" exhibited at CIAM 7 (Tyrwhitt 1949c: 1).

Tyrwhitt offered this critique against the backdrop of the work then underway of the Schuster Committee on the Qualifications of Planners, set up by the Ministry of Town and Country Planning in May 1948. Seeking to guide the universities that were building curricula to train planners, the Committee reconciled competing views of planner qualifications: the technical, design-oriented perspective and the generalist, synoptic view (Presthus: 1951: 43–42)—such as Rowse and Tyrwhitt had pioneered at the School of Planning. While the Schuster Committee did not publish its report until October 1950, it was clear that in the statutory system created by the Town and Country Planning Act of 1947, planning was "primarily a social and economic activity limited but not determined by the technical possibilities of design" (Ministry 1950: 69, 20).

In this context, Tyrwhitt was concerned about future prospects for Rowse and his vision for the School of Planning (as a "sort of 'post-post-graduate' training ... for 'planning underdeveloped areas of the world'"), but her own interests were closely aligned with the graduate program in civic design that Gordon Stephenson (1908–1997) established at Liverpool University in 1948. Tyrwhitt visited Stephenson in November 1949 to take a look at his program and "talk about planning education and its future generally." [T.32] She would have welcomed a job there. Tyrwhitt felt that "A field of practical planning in which all schools are weak is that of the sort of Civic Design that can evolve from an honest survey and an intelligent analysis." [T.3]

At this same time, Tyrwhitt's new partner, Wells Coates, wrote to Giedion on behalf of the MARS group formally proposing civic design as the theme for CIAM 8. Giedion fully agreed with this proposal, especially since it included a comparative historical component. He sent Tyrwhitt a copy of his reply confirming: "I have been doing for year what you proposed in my seminars on civic centers and social life in ancient, medieval, renaissance, and modern times: historical analysis according to our comparative CIAM methods." [SG.29] Through her rapport with Giedion Tyrwhitt quietly amplified the strength of the MARS group's bid to set the agenda and organize the next CIAM congress; Giedion, in turn, championed their proposal to Sert, who concurred. But Sert stressed that the MARS group would have to link its proposal to the theme suggested at Bergamo: *L'Charte de L'Habitat*. Tyrwhitt helped formulate this link through her translations of Le Corbusier's speeches— which he felt were *parfaite*—and her reports of his commission from Bergamo on the one hand, and her work with the MARS group defining the civic center as "the heart of the city" on the other. [SG.30, 31]

Giedion said that he came to support "the Civic Center after having seen in connection with our publication [on Bergamo] how important it would be that CIAM establish social outlines." [L.1] This publication fell through, however, due to lack of publisher interest. Nevertheless, Tyrwhitt continued working on the translations, which she felt "will be worth while in the long run certainly." [SG.32] In fact, Tyrwhitt's report provides the only coherent record of CIAM 7, and both Giedion and Sert drew on this material over the years. In the short-term her report served as the impetus for producing a CIAM book on town planning and civic design in advance of CIAM 8.

Uncertainty

Assuming that her contract with Vassar was set, Tyrwhitt booked passage to the US in late January 1950. Her plan was to work at Vassar and with Giedion at MIT until summer. She then hoped to guide the European portion of a housing and town planning study tour that Abrams would lead. Tyrwhitt also intended to look into a job with the UN that both Bauer and Carter had told her about the previous summer, likely with the new program of housing and town and country planning that the UN Department of Social Affairs had launched in June. Tyrwhitt was ambivalent about working for the UN; she was seriously interested in the job "in principle—but not seriously interested in living at Lake Success," a suburb of New York, where the UN was temporarily located. [B.10]

At age 44, Tyrwhitt was both free and forced to join the trans-Atlantic migration of European intellectuals to the US. She told Forrester (now Fifth Earl of Veralum): "This visit is not an emigration, but it's the only place that has offered me a job in the immediate future! That's a bit of an exaggeration as I've not been applying for any of the statutory jobs—I don't think I should be a good civil servant and intend to remain a free lance as long as I can possibly keep alive as such." Tyrwhitt had applied for "a position on the staff at Cambridge if the new planning school does get started there as part of the University." The attraction was a "job that keeps one's brain alert and pays one enough to cover rent and basic food. I've not found it awfully funny living on air—or catch as catch can—just recently, but it's not easy to find a steady income that does not bind one down intolerably—or else absorb body and soul ... like APRR." [T.33]

Depressing news about APRR's finances, rather than her own, had prompted Tyrwhitt to write Forrester. "APRR should ... I think be quietly suppressed," she advised:

> It did its job when the school was out of the question, and now had better pack up. The publishing side MIGHT have developed ... and the consultancy side MIGHT have developed ... but now that money is so tight for 'research' it's too difficult to maintain a middle path between government supported research (universities will always get these funds before us) and private consultancies. [T.33]

Forrester agreed and APRR's March 1950 Information Bulletin would be the final one, and it would feature the blurb for the Textbook as the lead article.

In early December, Tyrwhitt learned that the Vassar job had fallen through. Then the position at Cambridge fizzled. Tyrwhitt had been torn about that job anyway, as it would have meant another long separation from Giedion. Whether or not their relationship was romantic, Tyrwhitt had romanticized the idea of it and was clinging to that idea. At year's end she made it clear to Giedion that she wanted above all to help with his work:

> If I cannot be with you I must earn my living as I am best able ... . If I can be with you, then I can concentrate on doing the things I really want to do: help you, think out some things for myself, learn some more, write a bit – and take in rather than give out! ... I have no 'career' ambitions, and never have had. .... I am not exaggerating when I say that [your work] ... opened a door to the garden around my own house of thought, a garden that I had never explored before at all, though I had always been dimly conscious of its existence. I 'need' you for the development of my whole life far more than you 'need' me for companionship. [SG.33]

Representation of New Ideas and Possibilities

With time on her hands, while living in Ireland, Tyrwhitt offered to translate Giedion's MIT lectures, which he was using to develop his ideas for his next book. Giedion gladly accepted this offer. Tyrwhitt enjoyed translating the first text Giedion sent her, in German, on ancient Egyptian art and architecture, even though she did not understand his "exact meaning in all cases," which prompted her to ask to see what he wrote in English as well." [SG.34] To make Giedion's writing understandable in English was not just a matter of "picking words;" Tyrwhitt helped him clarify and conceptualize, which sharpened her own thinking. She found his lecture about how the vertical became dominant in art and architecture (replacing the multidirectional space conception of prehistory) particularly interesting, as she had been thinking along similar lines:

> This first started with my early studies of botany, but I only realized that it might be the basis ... of a philosophy ... when I hear Buckminster Fuller expounding on the fallacies of the cube. ... Perhaps you will miss out this entirely and come to the end question of the recovery of a roving eye—not conditioned by the vertical. [SG.35]

Ultimately, Tyrwhitt translated and edited a large portion of Giedion's MIT lectures. She was hurt, however, when he did not acknowledge her help. By way of apology, he wrote: "I feel very unhappy that I didn't even thank you immediately for the corrections. ... You know that normally lectures do not cost much energy but these five lectures were difficult, perhaps also because I liked to combine them with clarified thinking."

Concurrently, Tyrwhitt was helping Giedion with a series of articles to be published in *Picture Post* based on excerpts from *MTC*, an important opportunity to reach a broader public. After working with Tyrwhitt on the layout of the first article, on mechanization of the kitchen, the *Picture Post* editor invited her to write the text and find the illustrations for a second article, on the history of the bath (published July 1, 1950). This marked the beginning of a new dimension of Tyrwhitt's work with Giedion; she found the right imagery to visualize those ideas.

Festival of Britain: Town Planning Exhibition

Tyrwhitt's efforts to represent Giedion's ideas visually and verbally informed her own work at this time: preparing a scheme for the town planning exhibition at the Festival of Britain. She was hired in late January as part of a team that included an architect and a display designer. Her role was to write the script, and select and organize the presentation of visual material. The deadline for her first draft was the end of March, and she searched for images for the exhibition and the article on the bath at the same time. The objective of the exhibition was to represent "the way town planning can contribute to a better life for more people." The goal was to interest the general public, who after being "concerned for so long with programs for a better future … have naturally become somewhat skeptical of 'plans.'" [T.34] "I have been determined that the exhibition shall show actual achievement and not vague hopes … . If one can see real things then one can demand more and better examples of them!" [SG.37] To that end the Town Planning Pavilion was located at the entry to the Exhibition of Live Architecture in East London.

Tyrwhitt's scenario for the exhibition provided a tangible example of her argument that the purpose of planning is to promote the fuller development of people, beginning with an understanding of the full spectrum of their needs. Visitors to the exhibit proceeded through a series of bays that depicted a single town "as it relates to the lives of eight typical inhabitants: a baby, a schoolchild, an industrial trainee, a young married woman, a factory worker, an office worker, and an elderly couple." Displays show members of these groups taking part in "their Working Life, their Home Life, their Social Life, their Private and Personal Life as individuals … their Civic Life … and some form of Outdoor Life." After this example of "how pleasant life could be in a well-planned town," the exhibition demonstrated work in progress in Britain, especially the 14 New Towns then under construction. Visitors could then go into a technical exhibit or proceed to the exit, where a large diorama depicted "The Heart of the Town"—the theme proposed by the MARS group for the concurrent 8th CIAM congress—"the focus of social activities, an essential part of a healthy community." [T.34]

Another Atlantic Crossing

Once she had submitted the draft script to the Festival committee, Tyrwhitt was determined to travel to the US—her third trans-Atlantic crossing in two years—despite not having any jobs lined up to pay her way. She finagled a visa, and arrived in New York in April with only $25. Just shy of her 45th birthday, Tyrwhitt's leap of faith set in motion a chain of events that propelled her into the next chapter of her life: as an academic, a member of the CIAM inner circle and editor/translator of CIAM publications, and a consultant to the UN. She was in the right place at the right time with the right skills, and she knew the right people.

Giedion allowed Tyrwhitt to stay with him in Cambridge for a while. "But the iron curtain have to be especially careful manufactured for this case," he cautioned. He advised Tyrwhitt to tell Bauer her excuse for being in Cambridge was "the urgent task of the CIAM book … which is waiting for production & I have no time here to assemble & correct the English manuscript." [SG.38] Tyrwhitt in fact took over production of this book, *A Decade of New Architecture* (1951)—herein after *Decade*—which Giedion had nearly abandoned out of frustration with contributors. [L.2]

Shortly after her arrival Tyrwhitt's chronic cough became acute and she was hospitalized for a week. Abrams came to Tyrwhitt's aid; he commissioned her to write the introductory article for an issue of the UN's *Housing and Town Planning*

Bulletin that he was editing on community facilities and services in large-scale housing projects. This commission paid for Tyrwhitt's hospital bill, but, more importantly, it marked the beginning of her long and productive association with Ernest Weissmann, recently appointed Assistant Director of the UN Bureau of Social Affairs, in charge of the housing, building, and planning branch.

Tyrwhitt cultivated her connections and parlayed her transitional status. In May, en route between Cambridge and New York to meet Weissmann and Abrams about the UN article, she stopped in New Haven to see Christopher Tunnard (1910–1979)—a Canadian born landscape architect who had studied and practiced in Britain before emigrating to the US—who was then teaching city planning at Yale. This meeting led to an offer that Tyrwhitt accepted to be a Visiting Professor at Yale for the 1951 spring term. In New York Tyrwhitt conferred with Sert and Giedion on CIAM affairs.

Giedion no doubt encouraged Sert to use CIAM publication funds to hire Tyrwhitt to help produce a book on CIAM town planning. While Sert intended to edit the book, it would clearly build on Tyrwhitt's report on CIAM 7 and the town planning section in *Decade*. Tyrwhitt attended a CIAM Council meeting at Sert's New York office in June 1950 where her contract was approved, and the program for CIAM 8 also was discussed. In advance of that meeting, Sert had asked Tyrwhitt and Alberto Iriarte, a member of his staff, to study how to integrate the two themes proposed for CIAM 8: the further development of principles for "*l'habitat*" based on the ASCORAL grid, and the CORE (a term the MARS group used to refer to "civic centers" or "the heart of the city.") The CIAM Council approved of the grid as modified by Tyrwhitt and Iriarte and authorized Tyrwhitt to present it to the MARS group upon her return to London. In July she obtained the approval of MARS Executive Committee, too. In this way she played an essential role in the operations of the CIAM Council, which was straining to sustain the activities of the newly revived, significantly enlarged, widely dispersed international organization.

Subsequently, Tyrwhitt produced *Decade* while also playing the lead role in developing the contents of the new CIAM book on town planning, organizing CIAM 8, writing the article for the UN, and preparing her Yale lectures. The ideas that Tyrwhitt developed through her engagement with these inter-related activities and networks of people and institutions culminated in the discussions at CIAM 8, and in the book based on it, *The Heart of the City* (1952).

The Valley Section, Patrick Geddes' World Image

Significantly, the larger intellectual context in which Tyrwhitt undertook those new projects was provided by her continued excavation of Geddes's writings. She had already sparked a new round of interest in Geddes's ideas by the recent publication of APRR's *Town and County Planning Textbook* (1950b) and her abridged edition of *Cities in Evolution* (1949), which compounded the impact of her earlier edited collection, *Patrick Geddes in India* (1947a). In April 1950, upon learning from Bauer that Indian Prime Minister Nehru was interested in that book, Tyrwhitt sent him a copy. [B.11] Mumford used the publication of Tyrwhitt's edition of *Cities in Evolution* "as the occasion to write a reassessment of Geddes's contribution to city planning and civic philosophy" (1950: 81). Mumford encouraged Tyrwhitt to continue to curate and publish Geddes's writings.

Tyrwhitt wrote "The Valley Section, Patrick Geddes' World Image" in the fall of 1950. She intended to clarify the meaning of two of Geddes's phrases that had "become commonly used in planning circles: 'The Valley Section'— mainly employed by the geographer-planner—and 'Place, Work, Folk'—mainly employed by the sociologist planner" (1951a: 61). In Tyrwhitt's view: "Geddes's real contribution to planning thought and practice was to link these two concepts indissolubly with each other and with [Auguste] Comte's theory of 'Peoples and Chiefs, Intellectuals and Emotionals.'" To make this point she researched Geddes's writings and identified relevant extracts.

As in her previous Geddes books, Tyrwhitt didn't say much in "The Valley Section" in her own words; she carefully composed a narrative of her edited selections of his words. This constructed narrative presented a significant counterpoint to prevailing CIAM discourse on the value and use of the CIAM grid as an analytic tool, and the importance of the core, or heart of the city. It was published in *Journal of the Town Planning Institute* in January 1951, strategically timed to influence discussions at CIAM 8 and the book on CIAM town planning. But while some young architects and planners embraced the fertile ideas of Geddes as a humanistic alternative to CIAM's post-war positions on architecture and urbanism (Welter 2002), Tyrwhitt's intention was to enrich CIAM discourse, not subvert it.

Tyrwhitt (1951a: 61, 63) began with Geddes's notion of ecological economics, based on his 1881 proposal for a universal system for classifying social statistics. This system, grounded in biological principles, would make it "as easy to monograph a city or a village; a nation, a single household or even an individual, and to compare the facts." Matter and energy utilization could be measured at any level of complexity of organization—as a metabolism. In order to account for the functional/occupational as well as structural/population characteristics of a society, Geddes visualized how typical ways of life evolved in relation to a typical "stretch of landscape from sea to hill top." This panorama helped explain the dynamic interaction of humans and nature.

The watershed was "the essential unit" for a systematic, comparative study of cities and civilizations. But the student needed some "simplification of our ideas of history comparable to that of our geography," he wrote in 1904. The solution is a vertical section of the river system—a valley section—displaying dormant heritage: "not only does the main series of active cities display traces of all the past phases of evolution, but beside this lie fossils or linger survivals of almost every preceding phase" (Tyrwhitt 1951a: 63–64).

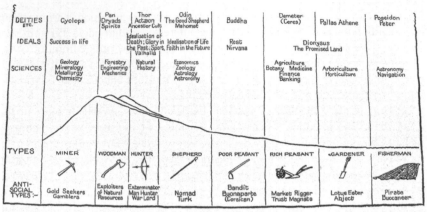

The valley section with rustic types.

Figure 11.3   Tyrwhitt researched Geddes's texts to explain how his concepts of the Valley Section and the triad Place–Work–Folk were inextricably connected

A Civic Survey organized according to this framework can "indicate the practicable alternatives and to select and define from these the lines of development of the legitimate EU-TOPIA possible in the actual city and characteristic of it; obviously therefore a very different thing from a vague U-TOPIA, concretely realizable nowhere." Tyrwhitt/Geddes reiterated: "not only that every scientific survey involves a geographic and historic exploration of origins but that the still unwritten chapters—the far-seeing glance forward, idealistic yet also critical—can be prepared by habitually imaging the course of evolution in the past" (Tyrwhitt 1951a: 64–65).

The key is how to interpret the facts assembled by such a Civic Survey in order to discern the EU-TOPIAN potentialities within a place. Therein lay the value of Geddes's Notation of Life diagram; it was his "thinking tool," visualizing the complex evolutionary expression and development of the modern city (Tyrwhitt 1951a: 65–66):

"We start with the general notion of a town ....We next see this simple town developing an experience and tradition in relation to its locality, its industry, its population and each of these in ... 'schools'" of thought at different levels: lore, learning and law. Eventually, there emerges a detached cloistered world, where the philosopher, scholar, scientist and artist articulate "new ideals, new ideas, new images." This Cloister "sooner or later projects its subjective world into reality." A transformation occurs when town, school, and cloister blossom together: "Thus ... only can arise the City proper [EUTOPIA], the renewed ideals of the Cloister creating or reforming social organization; the new ideas creating or transforming culture; the new imagery expressing itself in public art."

Tyrwhitt offered Geddes' diagrams as a comprehensive conceptualization of the dynamics of urban ecosystems, encompassing both temporal/objective and spiritual/subjective activities. The CIAM Grid, in contrast, was essentially a framework for organizing decision making in the planning and design process and presenting projects.

In addition, Tyrwhitt was rescuing the word "utopian" from its derogatory connotation; although Geddes invented the term "eu-topian" to distinguish a realistic/realizable image of an ideal community, grounded in a critique of the status quo, from an unrealistic/unrealizable "u-topian" fantasy. In doing so she reinforced the message of philosopher Martin Buber's book *Paths in Utopia*, published in London in 1949, which also called for utopian realism in planning and designing the post-war world.[3]

A Decade of New Architecture

Tyrwhitt wrote her Valley Section article while working on *Decade*. Giedion acknowledged that she brought "the book in a final form" (1951a: iv). However, that doesn't fully capture her creative contribution to this book, considered an important record of the post-war history of CIAM (Georgiadus 1993). Giedion wrote Tyrwhitt: "I was only furious that they printed my name all over the place ... It is *not an SG book!*" [SG.41]

The first part of *Decade*, Post-War Activity, emphasized CIAM's growing concern with community planning and "the need for new community patterns" at all scales (Giedion 1951a: 20, 25). That section celebrated both the diversity of ideas among CIAM groups as well as their cross-fertilization through collaborative work. Tyrwhitt's translation recast the CIAM grid as a "system of comparative visualization of a complex of rather difficult town planning problems in a form which enables not only planners but also a layman to get insight into urban design." Production delays enabled the first section of *Decade* to conclude with an excerpt from Tyrwhitt's report on CIAM 8—which recalls her rendering of Geddes's Notation of Life diagram:

> Urbanism is the framework within which Architecture and other plastic arts must be integrated to perform once more a social function....[This] **synthesis of the plastic arts can be most effectively accomplished in city 'cores.'**...The core ... is the expression of the collective mind and spirit of the community, which humanizes and gives meaning and form to the city .... (Giedion 1951a: 39)

---

3   Art critic Herbert Read had earlier called planners' attention to Buber's philosophy. In his Introduction to Gutkind's *Creative Demobilization*—which includes Tyrwhitt's APRR Broadsheet on Consumption of Fresh Food—art critic Read (1943: xiii, xv–xvi), praised the book as a contribution to Karl Mannheim's concept of "planning for freedom," and recommended that planners learn from Buber's advice that "the antithesis of constraint is not freedom, but unitedness."

Decade became, in Tyrwhitt's hands, a companion to *The heart of the city*, demonstrating how CIAM's post war work had evolved toward humanistic urban design. Tyrwhitt's influence is evident in the section on Town Planning [*Urbanisme*] that concludes "the illustrated part" of the book. Her "translation" of the text recapitulates her critique of the "The Neighborhood Unit" and the argument she presented more fully at CIAM 8: "The new city will have no sharply defined limits, yet it will have far more differentiated parts which are related to one another and yet freely assembled in starlike constellations." The text also cites Tyrwhitt's statement—from the *Town and Country Planning Textbook* (1950b)— referring to Geddes as "the first ... to coin the word 'conurbation' meaning 'to apprehend the whole as well as the part in the fourfold interrelation of regional survey, rural development, town planning and city design" (Giedion 1951a: 217).

CIAM Town Planning Book

By mid-July 1950, Sert had asked Tyrwhitt to take over the job of collecting all the material for the CIAM town planning book, since "the majority of contributions should come from Europe, anyhow." In the event that a lack of response from CIAM members made it impossible to produce the book, Sert suggested that they "do a smaller book on the CORE of the city which would summarize the work of the next Congress." [L.3] This could include a contribution from Giedion on the historical background of city centers and a section related to the arts, which developed the discussion at Bergamo. Giedion agreed it was better for Sert to focus on one book, on civic centers. "Books need time ... especially books on urbanism." [SG.40]

Tyrwhitt, however, objected. "I think it a pity to finalise a book on the CORE before we have had the 8th Congress," she told Sert. She proposed producing the book in two parts: "one half to be ready at the 8th Congress and the other to be prepared ... very quickly after it." Meanwhile, she collected as much material as possible in visits to Amsterdam and Paris. Sert relied on her judgment, as he was busy working in South America. He looked forward to working with Tyrwhitt when she would be at Yale, and hoped Giedion would be returning to MIT in the spring. "It will be easier if the three of us are together here," he wrote, signaling the crystallization of this key triumvirate. [L.4, 5]

Meanwhile, Tyrwhitt was living precariously on limited funds, and had to borrow money even as she volunteered to do more work for Giedion. Luckily, she was hired to oversee completion of the Festival of Britain town-planning exhibition. This job paid enough to allow her to stay in London through the autumn. Finally, Giedion told her to send him a bill for her work on *Decade*, which CIAM would pay. "You should not have worked again for nothing," he wrote. [SG.40]

In early November, Sert wrote to Tyrwhitt and Giedion urging that, due to lack of cooperation from CIAM members, they switch to the book on civic centers. "The title could be 'The Core of the Community' or 'Centers of Community

Life.'" For illustrations they could look beyond the work of CIAM members. But he needed Tyrwhitt's help in New York. "If you can bring as much material as possible from England and other European countries, and work with us regularly while in the US, then I think it could be done." His idea was to present a nearly finalized dummy of the book to the Congress in July, "leaving some space for additions that the Congress would advise." Giedion assured Sert that he agreed with "every point." [L.6, 7]

Tyrwhitt took a long time to reply, in part because she had been in Wales helping her sister with a new baby. But she may also have been reluctant to negotiate new terms for this work. After briefing Sert on her schedule and the material she would bring with her she set some personal limits in a post script: "PS. I've not actually said I'd like to work with you on the new book—but you know this is so, and I'm sure some fixed times can be arranged." [L.8]

Chapter 12
The Core of the Community

The Heart of the City and the Urban Constellation

In January 1951 Tyrwhitt arrived in New York with a bad cough—as she had the previous spring—but once she settled into her new routine at Yale University she thrived. She taught a seminar entitled "Origins and Development of Certain Planning Concepts" and a Civic Design Studio with Tunnard. She enjoyed her students, her colleagues, and campus life. She was still living a transient existence, but she was happier and more confident than she had been in some time. One day a week she took the train to New York to work with Sert on the CIAM book on "the heart of the city." By February she had a draft synopsis ready to circulate to potential contributors.

Tyrwhitt also continued to work long-distance for Giedion, who was writing new chapters for the third edition of *STA* on the development of certain urban planning concepts in medieval and Renaissance Italy; Tyrwhitt was talking in her seminar about those same topics. Under deadline pressure Giedion also sent Tyrwhitt his new chapter on Mies van der Rohe to translate. "It is such a difficult time absorbing task, because each word is weighed to the extreme … . It is a dance to get the meaning without nailing down or simplifying the sense. … so if you will try, try." [T.35] Ultimately, Tyrwhitt translated or "corrected" all of the additional text in the greatly expanded third edition of *STA* (1954).

Tyrwhitt could not afford to relax in her temporary position at Yale so she capitalized on her extensive connections to organize a lecture tour during the mid-term break. Stops in Chicago, Los Angeles, San Francisco, Seattle and Vancouver, British Columbia, served multiple purposes: find her next job; and earn enough money to devote two weeks before CIAM 8 to help organize it, and two weeks afterwards to pull together materials for the book. Sert asked her to bring back information about the CIAM groups in California and who would attend the Congress.

Equally important, on Tyrwhitt's lecture tour she renewed old acquaintances and cultivated new relationships that led to important collaborations. She forged an enduring bond with Meyerson, then fighting to save the planning program at the University of Chicago, and stayed at the home of Isaacs, who was commuting between Chicago and Harvard, where he was teaching with Gropius. She gave a talk on the CORE for students at the Institute of Design, where Chermayeff was in his last days as director. Fritz Gutheim called her in Chicago from Washington to ask her to moderate a panel—Albert Mayer and Clarence Stein were to be the speakers—at the American Institute of Architects (AIA) conference there in May. Gropius was then urging Sert to invite more planners to join the US CIAM group. Sert agreed. When Tyrwhitt returned to Chicago Sert asked her to invite Isaacs and sociologist

Louis Wirth to come to CIAM 8. On her own initiative she invited Meyerson, who was going to be in London at that time. In Los Angeles she was the guest of architect Richard Neutra. In San Francisco, she caught up with Bauer, then teaching at the University of California, Berkeley. Peter Oberlander, a recent Harvard graduate invited her to the University of British Columbia, where he picked her brain to strengthen the new planning program he had started there. [T.36]

Amid the percolation of ideas stimulated by Tyrwhitt's teaching, travels, lectures, and work on the CIAM 8 book, a visit to György Kepes (1906–2001) in Cambridge to collect illustrations,[1] yielded an important analytic insight: a further development of Geddes's concept of the conurbation. Kepes, the former Bauhaus teacher who had started a program in visual design at MIT in 1947, gave Tyrwhitt a tour of his exhibition, The New Landscape, for which he assembled scientific images that were made with new visualization technologies such as x-ray machines and infrared sensors. Tyrwhitt recalled that the "photographs of the heavenly constellations ... of microscopic biological life ... of plant cells ... of whirlpools and deserts ... of inorganic crystalline formations" inspired her to come up with the concept of "the urban constellation" to describe the dynamic relationships of cities, villages and towns, organized around "a vital city center. ... Both in the pictures of organic life and of inorganic matter—whether on the scale of the universe or of the molecule—one could discern a subtle orientation of apparently independent units towards a nucleus." [T.37]

Tyrwhitt tried out her new formulation at the AIA conference in May. Her talk, "The Next Phase in City Growth—The Urban Constellation," was well received. Burchard wrote a complimentary note: "it was a very high level and meaningful constellation of remarks—far more meaningful than anything anyone else said." [T.38] An excerpt of her speech was published in *Progressive Architecture* as part of a discussion of whether the threat of nuclear war called for urban decentralization. In response to the question, "Do new towns provide safety?" Tyrwhitt (1951b: 77) answered a resounding, "No. There must be a vital city center to which all parts of the [urban] constellation have access. ... Only in a living space that contains within it sufficient diversity of opportunity, can the human spirit gain that confidence and resilience that enables it to develop its full potential—and even at times to 'rise above itself.' The only defense against death is life."

Further invitations followed, notably to Toronto, where she explored opportunities for her and Coates, who had grown up in Canada. But at the end of her term at Yale Tyrwhitt had "no definite offers, only 'feelers.'" She returned to CIAM work, and took on the translation of Le Corbusier's article for the CIAM book. Giedion chided her: "Jacky, Jacky. Please don't do too much ... I don't agree with your supercharged schedule." [T.39] "That is the way I earn my bread and butter," she explained:

---

1   She was collecting images that Kepes produced to help Giedion's seminar at MIT in spring 1950 "find a way of representation both simple but not trivial" for what they wanted to say on "civic centers and social life." [SG.36]

If one is to work in a profession that is mainly men, one has to be better at some parts of it at least—and the men don't resent superiority in routine efficiency. They do resent any show of superiority in mental capacity. The life of quietness has never been possible for me—I could never get paid by anyone to be quiet. [T.36]

She also felt well on her way, though, to writing what she called a "me—book—based on her seminar, but to do so she would have to resist the tendency to subordinate her own work to other people's priorities; that continued to be a major challenge.

CIAM 8

Tyrwhitt returned to London in late June 1951 to help prepare CIAM 8. The congress took place in early July at a venue that offered a bucolic setting with few distractions, but also easy access for visits to New Towns and the Festival of Britain, which featured work by many MARS members, including Tyrwhitt. Tyrwhitt had co-authored the MARS group's definition of the Core: "the element which makes a community a community, and not merely an aggregate of individuals;" and she and Coates had produced the program inviting members to study the Core at five "scale levels" of community: housing group, neighborhood, town or city sector, city, and metropolis. Participants presented their work formatted according to the MARS version of the grid which she helped devise, and she arranged the grids by "scale-level" for discussion.

Tyrwhitt introduced her concept of the urban constellation—"a new term to the planning dictionary"—as an organizing principle for the five scale levels of community in her opening remarks as chair of the session on the Social and Historical Background of the Core. "This urban relationship ... only functions when there is a vital city center to which all parts of the constellation have access." [T.37] Tyrwhitt followed up her remarks during a roundtable discussion, observing that the vitality of the city center depended on informal social interaction and the presence of "a great diversity of people." In order to generate and support such vitality "there should be a close physical nearness of all that goes to build up the heart of a city. There should be no distinct zones of separation between, for instance, the shopping centre, the cultural centre, the administrative center" (Conversation 1952: 39). These comments posed a critique of the formalistic designs presented by Le Corbusier and Sert, if not their texts.

The Heart of the City: Towards the Humanization of Urban Life

Tyrwhitt's remarks at CIAM 8 were limited but she was able to secure a place in the record of CIAM discourse for her conception of modern urban planning and design through the key rapporteur role she played, and in producing the book based on it: *The Heart of the City: Towards the Humanization of Urban Life* (1952), for which she was credited as translator as well as lead editor, along with Sert and Ernesto Rogers. Tyrwhitt shared responsibility with Rogers for supervising the

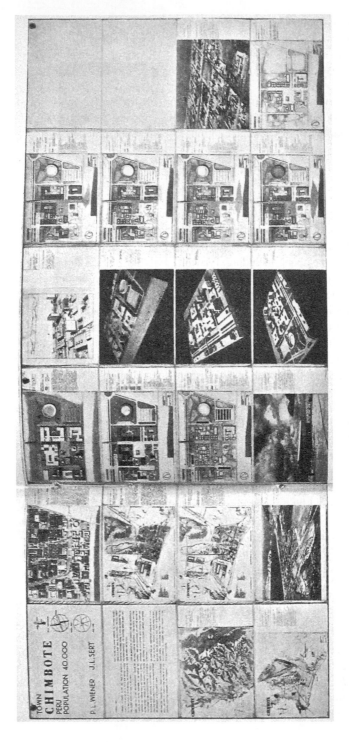

Figure 12.1 The MARS Grid, as illustrated in *The Heart of the City* (1952), provided a standardized presentation format to facilitate comparison of diverse projects

book's publication in Europe; Lund-Humphries in London published the English edition, and Hoepli in Milan published the French, Italian, and Spanish versions. She was uniquely qualified to play this role because of her extensive experience producing publications, her relationship with Lund-Humphries from her APRR years, her language skills, and her mobility. She was not only free to travel, but with her relationship with Giedion hidden behind an iron curtain, she moved easily among the CIAM membership, once a close knit group of friends, but now riven by personal, cultural, and generational conflicts.

By virtue of her position at the center of this international publication process— distance from Sert allowed her some maneuvering room—Tyrwhitt could arbitrate the book's contents. She wisely used her position, though, to solidify her relationship with Sert as a trusted colleague. For example, shortly after the congress, Sert sent Tyrwhitt a list of his concerns about the resolutions that had been passed, among them that there was insufficient stress on the civic character or social importance of the heart of the city. "It will be difficult to translate the resolutions from French to English and I hope the translation will give us the opportunity to clarify some of these statements." [L.9] Sert was quite pleased with Tyrwhitt's translation and summary, which formed a conclusion to the book as well as to this era of CIAM. The Core as a means for the "animation of spontaneous nature ... seems a heritage that our group, after twenty years' work, can now hand on to the next generation. Our task has been to resolve the

Figure 12.2   The core of the city was above all a place for spontaneous social interaction

first cycle of the work of CIAM by finding a means to *transform the passive individual in society into an active participant of social life*" (Short 1952: 168).

Tyrwhitt did not assign herself a chapter; she hoped to expand on her ideas in her introduction to the book's second part (examples of projects). Due to space constraints, however, her text had to be brief. Tyrwhitt used the limited space she had to emphasize (1952: 103–104): The Core was not "a group of civic buildings together with their related open spaces," but rather, "the gathering place of the people. … whether planned or not … a physical setting for the expression of collective emotion." The "cure for our … amorphous modern cities" was not to be found in urban decentralization along garden city lines but "by the creation of new Cores—new concentrations of activity-by a visual emphasis upon centers of integration rather than upon bands of separation." In this context the "task of the architect and planner is to appreciate the attributes of each Core and enable these to be developed so that the people of that community can derive the greatest benefit from coming together." This notion built on a talk she had earlier given at Harvard, where she said: "the most important thing about the Core is no-thing—is emptiness—a space that can be filled with human emotion." [T.40] That idea of the empty center—the space in-between persons where inter-actions take place— echoed the dialogic philosophy of Buber, presented in *I and Thou,* first published in English in 1937. In 1953 Buber spoke directly to architects (and planners) in his Foreword to EA Gutkind's *Community and Environment*: "The architects must be given the task to build for human contact, to build an environment which invites human meetings and centers which give these meetings meaning and render them productive" (1953: ix). Tyrwhitt embraced that idea.

CIAM 8 "heralded the final period of CIAM, in which it would concentrate more and more on social aspects of urban planning," Giedion declared (1967: 702). Tyrwhitt had a lot to do with steering CIAM in that new direction, both organizationally and intellectually. Giedion warned Tyrwhitt away, though, from doing further heavy lifting for CIAM. "It is an animal which devours its children, anyway those who serve it." [SG.42] But Tyrwhitt's hard work secured her place in CIAM's inner circle, where she found a sense of professional identity while other aspects of her life remained in flux; in that context her sacrifices for CIAM proved to be a good career move. In late summer, just as she was feeling overwhelmed and under-appreciated, she received a letter from Sert affirming: "You have done a wonderful job both in the report for CIAM and the book … You have practically given all your time to CIAM during the last two months and we have to work out some kind of arrangement for past and future work." [L.10]

Toronto as a Launch Pad

By then Tyrwhitt was already preparing for another move; she had accepted a position as Visiting Professor of Town Planning at the University of Toronto, starting in September 1951. While a Toronto base would make it easier to work with Sert in New York, Tyrwhitt would have to juggle her new job along with work on the CIAM book, which dragged on for more than nine months via a stream of correspondence and periodic European trips for face-to-face meetings.

Tyrwhitt was invited to the University of Toronto for one year—she was replacing Anthony Adamson, a locally prominent architect, who was on sabbatical—to set up a new graduate program in town and regional planning; she also gave a lecture course and supervised the work of the 5th year architecture students. At first Tyrwhitt was lonely in Toronto, where she lived in a small room in the Women's Club, and also ate most of her meals there as women were not allowed in the university commons! But she was glad to have an academic status, and from her base in Toronto she established herself in the North American academic community of urban and regional planners and designers through frequent travel, for example, to Pittsburgh, PA, to attend an ASPO congress in October.

In her first term Tyrwhitt repeated the lecture course that she had given at Yale, which focused on utopian traditions in town planning. She began with "utopias of the past, proceeded to "ideal new towns today," covered elements such as open space, traffic and shopping centers, examined the neighborhood unit in theory and practice, and concluded with centers of community life. [L.11] Tyrwhitt typed out her lectures; thinking they might be the basis of the town planning principles book she intended to write, and she sent them to Giedion and Sert for comments.

In the meantime, Giedion had encouraged her to seek out Marshall McLuhan (1911–1980), an English professor at University of Toronto, whom Giedion had met in St. Louis while researching *Mechanization Takes Command (MTC)*. McLuhan had sent Giedion a copy of his book, *The Mechanical Bride* (1951). "I estimate him as one of the few on the right way," Giedion told Tyrwhitt. "I wrote him: … one of my best friends The PROFESSOR of town planning … is living this year in Toronto. Please make contact with her. … Look him up Jacky." [SG.43] Tyrwhitt promptly met McLuhan and his wife Corinne. She greatly enjoyed their company and became a lifelong family friend.

At the end of the term, in what became an annual tradition, Tyrwhitt spent the Christmas holidays with the Abrams family in New York. There she fell into "a series of interlocking circles of people all of whom I have something in common with," she told Giedion. Her social success left her feeling "happy, relaxed and rather exhilarated," whereas in Toronto, she claimed to feel no connection, even to the people she liked best, including McLuhan. [SG.44] "The kind of thing I've been writing about interests NO*ONE at all," she maintained. In New York, however, she had seen Martin James, an editor of the innovative arts journal *Trans/Formation*—and translator of *MTC*—for whom she had promised to write something on the subject of "Ideal Cities and the City Ideal" the previous spring. Newly inspired, Tyrwhitt quickly wrote the article, based on her first two lectures, and sent it to Giedion for comments, as it "MAY be the first chapter of the book that I must get written before too long on a re-analysis of town planning ideals." [SG.45]

Ideal Cities and the City Ideal

In "Ideal Cities and the City Ideal" Tyrwhitt offered an overview of utopian proposals for new towns. She noted that the impulse to create such alternatives arose at times of crisis or revolution, and posited four revolutionary periods with corresponding utopian imagery: the revolution of learning associated with the Renaissance; religious awakenings that gave rise to sectarian communitarianism; industrial transformation during the early nineteenth century; and transportation in the twentieth century, which consisted of three parts: before, between, and after the world wars. In the post-war period she described an emergent "revolution of humanism," which "can be glimpsed between the lines of S. Giedion's *Mechanization Takes Command* (1948) and … run[s] like a gleaming thread through the CIAM discussions on the 'Heart of the City' (1952)."

Tyrwhitt's analysis of utopian planning traditions was a promising beginning for a larger project, but she did not develop it. She preferred to ally her work with Giedion's research on prehistoric art and architecture. She took advantage of the University's resources to do what she could to find literature or track down images to send him. She used Toronto's excellent Chinese library to study ancient Chinese art and architecture, "in case it is useful." That spring Giedion was investigating the origins of monumental architecture. Tyrwhitt wrote that she should soon "be able to say something on monumentality." [SG.46] While intellectually stimulating, this research was a form of escape—a distraction from her own work—a daydream of an ideal future with Giedion, perhaps in Amden, the alpine village where he was then buying land for a weekend house.

Yet she also became involved in "the Canadian scene," where she was one of only four professors of town planning, and attended their annual meeting in February in Montreal. The others were Harold Spence-Sales, a School of Planning alumnus who started the program at McGill University; Peter Oberlander; and Joseph Kostka, a former student of hers at the School who started the program at the University of Manitoba. Humphrey Carver, now well established at the Central (later Canadian) Mortgage and Housing Corporation (CMHC), provided the start-up funding for these university programs and research on urban development. This place "is a vacuum that is up to me to fill!" Tyrwhitt told Sert. [L.12] In January 1952 she told Rowse that she felt ready to "get about the staff and find a [inter-faculty] team that really could do something." [T.42]

One reason Tyrwhitt felt comfortable with her Canadian colleagues was, that aside from Oberlander, they all had ties to Rowse. Tyrwhitt felt a strong sense of loyalty to Rowse, and was using her connections with Weissmann at the UN to help him get work in Pakistan. Instead Rowse wanted to enlist Tyrwhitt in his own scheme to plan for Commonwealth Development Settlement in Canada:

> If we could get a pilot scheme going in each Dominion, we would be more than
> half way to convincing those who have the power to make things happen.…If you
> in Canada would be prepared to join in a collaborative system of such regional

development projects, located in all parts of the world, it might be possible to build up a working organization, which could make consultation profitable. [T.42]

Tyrwhitt replied: "I am sowing seeds as hard as I can in Ontario, and would certainly be prepared to co-operate with anyone you suggest." She felt she could "do more by remaining in a University job, which gives one security of rent and at the same time considerable freedom of movement with some assurance of 'objectivity' when dealing with people." [T.41] It's worth noting that Rowse's ambitious proposal was in line with the formation of multi-national consulting firms at that time, such as the one established by the Greek planner Constantinos Doxiadis in 1954, at the behest of multi-lateral aid programs and foundations, specifically to organize the resettlement of the thousands of people dislocated by the war, decolonization, and urbanization.

Habitat: CIAM and the UN

In May 1952 Tyrwhitt left Toronto without a new contract, but, assuming that she would return, she left her books and winter clothes behind and kept her room. She was going to Paris, where Giedion had scheduled a CIAM council meeting, but stopped first in New York to meet with Sert and Weissmann on inter-connected CIAM and UN concerns.

The UN was considering forming a Working Group on Low Cost Mass Housing (*Habitation pour le plus grand Nombre*). The impetus came from two CIAM members: the Swiss engineer Jean Jacques Honegger (1903–1985) and the Russian born French engineer Vladimer Bodiansky (1894–1966). It was an outgrowth of their involvement with CIAM Commission IV: Industrialized Building generally, as well as their work with the firm ATBAT-Afrique and French architect Michel Ecochard (1905–1985) in Morocco, which was aimed at re-housing residents of the *bidonvilles* (shanty towns). Bodiansky wanted Tyrwhitt to organize the working group, if the Economic and Social Council (ECOSOC) approved funding at its next session. [SG.47] Honneger and Bodiansky also proposed an exhibit of CIAM work on this theme to be held in Geneva in September, when there would be a meeting of the Committee on Housing, Building and Planning of the UNECE. [L.13]

Giedion was unaware of this initiative. He had called the May Council meeting to discuss the prospect of handing over leadership of CIAM to younger members, beginning with the upcoming ninth congress, which was to be organized by ASCORAL on the theme of *L'Charte de L'Habitat*. The Council met in Paris in May since most of the senior members were there anyway to discuss the new UNESCO headquarters, and they were unable to convene in June in Sigtuna, a town near Stockholm, where delegates would plan CIAM 9. Since Sert could not go to Paris, he briefed Tyrwhitt in New York so she could represent him. [L.14] Giedion was pessimistic about the leadership transition. He wanted, in particular, founding members Le Corbusier and Gropius to weigh in on the question: "Do we want to destroy CIAM or ... keep it?" [AG.48]

Tyrwhitt played a pivotal role at the Paris meeting by virtue of her now official capacity as secretary to the Council, her unofficial role as a confidant of Sert and Giedion, and her connection to the UN and the *Habitation pour le plus grand Nombre* project. She reported in a circular to CIAM delegates, also signed by Giedion and Le Corbusier, that the Council had made no progress on translating the word "habitat" into English, but it had "accepted that it meant something larger than 'housing' and smaller than 'neighborhood'—in other words, the setting of daily life." [L.15] Tyrwhitt informed Sert that it was impossible to change the CIAM 9 *Charte de l'Habitat* theme, which implied that he had wanted to. She reassured him, though, that Le Corbusier was not being dogmatic. He had announced that he no longer was "confident about the way men should live in this changing world. ... The 'Habitat' is clearly an element of living space ... but how it should be organized with the other elements is less and less clear." [L.16]

Back "home" in London, Tyrwhitt reported on the Paris meeting at a special MARS gathering to discuss CIAM 9; MARS members were dissatisfied with the agenda for the upcoming Sigtuna meeting. [L.17] They were also concerned that ASCORAL's proposed program for CIAM 9 seemed "confused and purposeless." Tyrwhitt suggested that the MARS group draft an alternative program based on its own interpretation of *"Charte de l'Habitat,"* which she suggested meant: "the dwelling and its immediate environment." Following considerable discussion about how to study this subject, and the meaning of *Charte de l'Habitat*, Tyrwhitt was appointed to a subcommittee to state the MARS viewpoint and develop an alternative program for Sigtuna. [T.43]

Tyrwhitt reported all this to Sert, along with her concern that Sigtuna was going to be a "fiasco." [L.17] Her letter prompted Sert to urge Giedion to attend Sigtuna, which he feared "may turn out very badly." [L.17, 18] But Giedion was busy on his book and also told Tyrwhitt that he wanted to see what would happen "when none of the bad boys, accused to have had since 10 years a false policy, will be there." [SG.49] The problem was not a simple generational cleavage within CIAM, though. Tyrwhitt warned Giedion that an ideological row was brewing within MARS: "one group for a thorough purge, another for quietly killing MARS, another for letting sleeping dogs lie." [SG.50] Giedion informed her "confidentially" about another quarrel: Le Corbusier was protesting the UN's invitation to Honegger and Bodiansky to organize the CIAM exhibition in Geneva in September. [SG.51]

Tyrwhitt represented the CIAM Council at the Sigtuna meeting and played a conciliatory role, downplaying the concerns of the MARS group and concentrating on getting decisions made. Bodiansky chaired a committee that worked out a preliminary grid that "leaves plenty of loopholes for initiative." All of the delegates now "know that there is no cut and dried view of what constitutes HABITAT, and it's up to CIAM 9 and the work that is brought there to define the matter," she told Sert. This meeting confirmed her belief "that the 'middle group' of CIAM had little to contribute and one must jump to the 'young.' The only thing is that the 'young' are not really ready to WORK," she advised [L.19].

Interlude: Demise of the School of Planning

While in London, Tyrwhitt also discussed the future of Rowse's School of Planning; Pepler, chair of the School's Board of Directors, was leading a last ditch reorganization effort to save it. "The re-establishment of the School after the War was due to you more than anyone else and we cannot be sufficiently grateful to you," he told Tyrwhitt. "Future prospects are at the moment grim," however. [T.44]

Tyrwhitt leveraged her connection to Weissmann to try to help the School, just as she tried to help Rowse personally. Students in the diploma course were mainly returning officers who held Ministry of Education demobilization grants, which would end in December 1952. The rest were graduates from the Colonies and Dominions who were funded by their governments. Tyrwhitt proposed to Weissmann that the School specialize in training planners for work in Commonwealth countries where there was a growing demand for development personnel. She also tried to negotiate an arrangement for students through the UN Fellowship program. The new Inter-American Centre for Research and Training in Bogota might serve as model for UN support for Rowse's School. Weissmann agreed: "It may be possible to initiate a number of such training centres serving various parts of the world; in this case the School of Planning ... could play an important role within this framework for Europe." Weissmann affirmed: "the Housing and Town and Country Planning Section would be very happy to do what it can in ... recommending your course on Commonwealth Development Problems once it has been established." Tyrwhitt then drafted a syllabus for a Special Course for UN Fellows, and Weissmann sent word that he wanted a copy of that syllabus sent to him directly in Geneva, where the Housing and Town Planning committee of UN-ECOSOC was convening. [T.45, 46] Nothing came of that initiative, though.

Tyrwhitt continued to do whatever she could "both to further the ideas of the School, and to help any individual associated with it," until it was officially closed in December 1952, when Department of Education grants ended. [T.47] Tyrwhitt's work on behalf of the School was selfless, but it undoubtedly reinforced her relationship with Weissmann, who was likely impressed by her loyalty, initiative, and ability to work efficiently trans-nationally through her extensive networks. In September Wiessmann appointed Tyrwhitt to the *Habitation pour le plus gran Nombre* project, which the UN had agreed to fund. She received a letter from Bodiansky with the news; he was named "rapporteur general," and she was responsible for overall organization. The position—full time for ten months, starting in May— would involve collecting evidence from experts in many parts of the world and convening a "technical working group meeting" in Morocco in the fall of 1953. Tyrwhitt was excited but cautious, knowing "how uncertain all these things are." [SG.47]

Toronto Year Two: Continuity and Change

Meanwhile, Tyrwhitt returned to Toronto, having signed on for another one-year position as Visiting Professor. She was becoming more established there. At the end of the fall term she proudly repeated to Giedion a colleague's compliment: "My advent there last year had been greeted with the deepest suspicion by both staff and students but that I had won 'all hearts.'" [SG.52] In January 1953 she was asked to assume more responsibilities when Adamson, who had won a local election, took a leave of absence. This meant she again represented Toronto at the annual meeting of Canadian planning professors.

In addition Tyrwhitt had agreed to be part of a team McLuhan formed to submit a proposal to the Ford Foundation; she saw this as a way of getting Giedion invited to the university during the next academic year, which would make her more willing to stay. McLuhan had initially informed Tyrwhitt that the Rockefeller Foundation wanted to fund a research center at the University of Toronto in commemoration of the communications scholar Harold Innis (1894–1952), who had died in November 1952. The way McLuhan talked about the proposed center to support the sorts of inter-disciplinary studies Innis had done sounded a lot to Tyrwhitt like Giedion's ideas for a Faculty of Inter-relations; she got McLuhan excited about the idea of inviting Giedion to be part of it. [SG.45] McLuhan then involved Tyrwhitt in the separate but related proposal to the Ford Foundation, to support an inter–disciplinary study of communications and culture inspired by both Innis and Giedion. [SG.46] Tyrwhitt was one of five faculty sponsors of the proposal, along with McLuhan (English), Edmund Carpenter (Anthropology), and others from Political Economy and Psychology. The proposal described her as a "pioneer of interdisciplinary studies in Britain ... long associated with the research projects of Siegfried Giedion," and who was advancing such cooperative efforts between town planning and several related departments at Toronto. [F.1]

Looking Beyond Toronto: Towards CIAM 9

Throughout her time in Toronto Tyrwhitt solidified her relationship with Sert through her work for CIAM, and by disseminating and clarifying CIAM ideas in her teaching, talks and publications. In November 1952 Sert asked for her help in composing a memo in advance of the next CIAM Council meeting—which neither of them could attend—to ensure that clear statements were incorporated in the minutes. [L.20] In her response Tyrwhitt called Sert's attention to the latest issue of *TEAM*, a newsletter edited by two young architects, announcing that the First Junior Congress of CIAM would be held in conjunction with CIAM 9. She urged him: "PLEASE take no action that would discourage this. Even if *TEAM* itself and some of the active youngsters seem a bit green please let them have freedom to try and work things out for themselves." She further advised against his recommendation that an "Advisory Committee" of senior members supervise the work of new CIAM groups:

> If we do not allow this open freedom the new generation of CIAM will consist entirely of 'followers'—and we already have enough of these. ... My own view is that the CIAM problem is not so much ... of spreading an existing credo; but ... the promotion of a creative sort of co-operation between the younger architects with really fertile imaginations. [L.21]

At the CIAM Council meeting in January 1953, after much discussion of the meaning of HABITAT, the Council agreed it should remain the theme of the ninth congress.

In January Tyrwhitt also learned that Sert had been named Dean of GSD and head of the Architecture Department, succeeding both Hudnut and Gropius, who had recently resigned. "How will it affect you?" she asked Giedion. "How will it affect me?" This news was unsettling. One voice in her head said "there is a good job to be done [in Toronto] IF ONE STAYS HERE and does the job thoroughly; another winces at the very idea ... unless one could get away every year." [SG.47] Tyrwhitt would have liked to (as Sert hinted might be possible) "help as a guest at Harvard." [L.22] Giedion thought that Sert's appointment was "a wonderful victory for CIAM." He planned on visiting the US in the spring and he would speak to Sert then. He told Tyrwhitt, though, he preferred that they not teach at the same school, "to make things easier for everybody." [SG.48]

But Tyrwhitt's future in Toronto remained uncertain, even though in her second year she had begun to "really feel things are being accomplished." The faculty senate had to decide whether or not to establish the planning course, rather than continue the experiment. In an effort to secure the program, Tyrwhitt told G.B. Langford, the chairman of the Graduate School Town Planning Committee: "to make things easier for them ... if the appointment of a woman is the stumbling block, here is a good opportunity for us to part amicably." [L.22]

She explained to Giedion: as "a female human being ... even in work one is more satisfied to assist—to help forward—the work of someone else than to sit in the chief seat oneself." (This was a leadership style she also admired in men, notably Gropius: who "works quietly behind people—gently inducing them to do things or behave in ways that will move matters forward, but always leaving it to THEM to act.") Tyrwhitt was also working "in a background capacity" to help Coates in Toronto. But she now doubted "it would ever be possible to work WITH Wells—he is not a 'co-operator.'" She complained: "Wells wants the thing to be <u>his</u> but finds it useful to use me as a source of information." Searching for personal satisfaction, Tyrwhitt studied German, and devoted her Saturdays to research for Giedion on "prehistoire." And she found fulfillment in her familiar role as family caregiver, assisting the McLuhans following the birth of their sixth child. "It's astonishing how unhelpful 'neighbours' are in situations beyond the first child ... and this is not unaccustomed ground for me." [SG.49, 50]

UN Technical Assistance to India

Weissmann telephoned Tyrwhitt in Toronto in early March 1953 to inform her that *Habitation pour le plus grand Nombre* had been postponed for a year because the political situation in Morocco was too volatile. However, there was a technical assistance job in India for which he wanted to nominate her. Tyrwhitt allowed Weissmann to do so, but said to Giedion: "Frankly I don't think the UN will want a woman." [SG.51] She was, though, extremely well qualified for the job.

Back in October 1951 when Tyrwhitt had attended the ASPO conference in Pittsburgh she had been recognized as a member of "a select group of planners" with foreign experience who were contacted by Philippine planner Antonio Kayanan, then assisting Weissmann, for their reactions to research the UN was conducting on neighborhood units. [L.23] Wiessmann had attended CIAM 8 and a concurrent IFHTP meeting to invite those organizations to contribute to that study and others being undertaken by the Housing and Town and Country Planning Section, including dwellings in tropical undeveloped areas and the education of planners. Those studies were part of an expanded program in housing, community planning and building, focused on the special needs of less developed countries, which the UN Technical Assistance Board (TAB) launched in July 1951 (Gardner-Medwin 1952).

Tyrwhitt had already exerted influence on UN technical assistance policy by then. In 1950, as a first step in preparation for the new technical assistance program, the UN dispatched a reconnaissance team, a Mission of Experts on Low Cost Housing led by Jacob Crane, to South and South East Asia. Crane's mission visited India, Pakistan, Malaya, Thailand, and Singapore; their report, published in July 1951, concluded that: "prevailing conditions in Asia create the greatest housing problem in the world" (Low Cost 1951: 11, 17, 67). Through her work with Weissmann that summer, as well as her discussions with Kayanan in October, Tyrwhitt likely knew that the Mission's report lauded a pilot scheme in Uttar Pradesh, India, which followed "the 'organic' process of village improvement preached 30 years earlier in India by Patrick Geddes," as a model for rural reconstruction that "may show the way for many others." The consultant to the government on that project was Albert Mayer, who, as noted previously, thought highly of Tyrwhitt's book, *Patrick Geddes in India*. The Report stated: "It is interesting to notice how nearly this individual, self-help approach corresponds to the Geddes method, and for this reason some extracts from his writings on Indian cities and villages"—from Tyrwhitt's book—were included in an appendix. The actions taken based on the recommendations of Crane's Mission of Experts set in motion a chain of events that took Tyrwhitt to India two years later.

Several of the Crane mission's recommendations were adopted by ECOSOC in July 1952, including a program of action for the benefit of low income groups in the developing world based on a comprehensive approach to housing and community improvement considered in relation to national economic development objectives (Resolutions 1952: 25). The UN Economic Commission for Asia and the Far East (ECAFE) subsequently set up an Inter-Secretariat Working Party on housing and building materials, which recommended greater use of local materials and more extensive housing and community improvement programs in the region. [T.48] ECAFE also discussed possibly assisting the Indian Government with an Exhibition on Low Cost Housing in the spring of 1952. The Government of India would have preferred cash, but accepted UN assistance in the form that was offered: an expert adviser, a contribution to the exhibition, and a regional seminar to stimulate further international exchange of information.

That was the state of affairs when, in late March 1953, the UN Technical Assistance Administration (TAA) invited Tyrwhitt to apply for the position of "Project Director." Her work would be: to advise the Government of India on the organization of an International Exhibition on low-cost housing emphasizing the use of local materials and skills, to be staged in New Delhi in early 1954; and to organize a concurrent regional seminar dealing with low cost tropical housing and to draft the report on it. Her work would begin in May or June and run for ten months. Tyrwhitt assured Giedion she would use the opportunity, if offered, for their collective benefit: "Of course the India seminar and exhibition could (and should) feature CIAM work." [SG.52]

Sert was thrilled at the prospect that materials assembled for CIAM 9 could serve as the basis for Tyrwhitt's exhibit in India. She spent several days in early May with Sert and Giedion finalizing the CIAM 9 program. In addition to clarifying the agenda, they also worked with material Weissmann had collected at the UN to devise a more useful grid than the one prepared by ASCORAL both to organize Tyrwhitt's exhibit, as well as for publication in a proposed book on "The Human Habitat." [L.24]

Passage to India

In May the Indian government accepted Tyrwhitt's appointment for both positions: to advise the Ministry of Housing on the Exhibition of Low Cost Housing; and to organize—the first—UN Seminar on Housing and Community Improvement in Asia and the Far East. Ironically, while being female presented an obstacle to her employment in academia, Tyrwhitt became the first woman to lead a UN technical assistance mission, a reflection of the new social spaces being created by transnational actors outside of—albeit connected to—traditional male dominated institutions.

Tyrwhitt began her assignment in May 1953, by spending a "grueling" month at UN headquarters in New York. [SG.53] Aside from helping decide which subjects would be discussed at the Seminar and which experts would be invited to lead the discussions, she learned how TAA projects were administered. She then went to London, where she arranged to meet with many of the people in England connected to India, to "get advice on who is who in the housing and town planning field, politically and professionally." [T.49] Among others, she spent time with Otto Koenigsburger (1909–1999), who had planned several refugee projects on behalf of the Indian Government. (She was already in correspondence with Fry who had been in India working on Chandigarh, a new state capital, along with Jane Drew, Pierre Jenneret, and Le Corbusier.) She was able to consolidate British expertise related to housing in India and channel it towards the UN at this critical initial stage in the technical assistance program. This type of face-to-face communication was essential as the UN cultivated a pool of experts to send out to difficult field assignments (Weissmann 1978).

Tyrwhitt also visited IFHTP headquarters, now located in The Hague, Netherlands. She faced "a very sticky situation vis-à-vis the IFHTP," which was holding its first regional conference for the countries of South-east Asia in conjunction with India's Exhibition of Low Cost Housing, immediately following the UN seminar, and involving many of the same people. [SG.53] Tyrwhitt was to coordinate the two seminars—and avoid conflicts between them. Some people ascribed the source of these conflicts to jealousy felt by Pepler, then 71, who in 1952 became Honorary President of IFHTP for life. [B.12] With her longstanding ties to Pepler and to IFHTP, Tyrwhitt was uniquely situated to mediate.

In Geneva, Tyrwhitt visited the headquarters of the four specialized agencies, UNESCO, International Labor Organization (ILO), World Health Organization (WHO), and the Food and Agricultural Organization (FAO), which played considerable parts both during the Seminar and as sponsors of the UN's contribution to the Exhibition.[2] In the course of those discussions Tyrwhitt developed her proposal for the UN's contribution to the Indian Government's Exhibition of Low Cost Housing: a model village that dramatized a housing and community improvement program in terms of daily village life. Based on discussions with WHO officials regarding environmental sanitation, she identified four areas to stress: water supply, latrine construction, ventilation, and food storage. Another official suggested including a maternity and child welfare center. ILO officials proposed some training workshops. She also identified related initiatives to tie into: an ILO regional conference on housing planned for September in Tokyo; an ECE conference planned for Geneva in October on Building, Civil Engineering and Public Works; and the formation of an International Council for Research and Documentation in Building. Her reconnaissance sorties and advice helped the Housing, Building and Planning Branch, under Weissmann's direction, to establish its role as coordinator of the projects of the regional commissions and specialized agencies that focused on human settlements.

Shortly after Tyrwhitt's June arrival in India, officials of the Ministry of Housing approved her preliminary proposal for the model village that would serve as a centerpiece of their Exhibition. Her next job was to write letters of invitation to all Governments in the region informing them of the Seminar and asking them to nominate participants. Those selected to lead discussions included: Crane, Abrams, Frederick Adams (Chair, MIT Department of Urban Planning) and Rafael Pico (Puerto Rican Planning Board) from the US; GA Atkinson, Building Research Station, UK (and liaison to the British Colonial Office); I.J.P. Thysse, Netherlands, Professor, University of Bandung, Indonesia; Arieh Sharon, head of

---

2   This account is based on JT.50–53.

the Government of Israel Planning Department; GF Middleton, Commonwealth Building Station, Australia; and Doxiadis, who was then based in Australia. [T.54]

Tyrwhitt lived initially in a guest house for members of parliament. Although her lodgings were near the commercial heart of "Lutyens' Delhi" she could see immediately that transport would be problematic. "This is a city of the sort of distances that Corbu is again manufacturing!" she told Giedion. [SG.54] But she barely had time to get used to her new surroundings. In early July Tyrwhitt went to France, to attend CIAM 9, after first making a stop in Israel.

In Israel Tyrwhitt found abundant examples of a synthesis of Geddesian and Bauhaus ideas being put into practice, inspired by Buber's philosophy, to solve problems posed by the arrival of waves of immigrants. Sharon had overseen much of this as the head of the department that prepared Israel's National Plan. He had also been involved in the planning and design of several kibbutzim—heralded by Geddes and Buber, among others, as a model confederation of autonomous cooperative communities. Middleton was in Israel too and explained his rammed earth building system. Tyrwhitt thought it would be useful in India and worth including in the UN's model village. At a TAA meeting Tyrwhitt met most of the experts, then working in Israel, who were good sources of information about housing suitable for hot-arid climates. [T.55] She was using her transnational connections to transfer technologies as well planning ideas. Ironically she was also following in her father's footsteps.

CIAM 9 Connection

CIAM 9, which took place during July 1953, in Aix en Provence, was poorly organized but well attended (Mumford 2000: 225–238). There were 40 "grids" displayed, most of which went with Sert to Harvard, where he planned to arrange an exhibition in the fall and would also select the material to be reproduced in the proposed book on Habitat. Weissmann was to bring to India those grids deemed suitable to exhibit there. Tyrwhitt advised that the most interesting grids from their point of view were from very diverse settings, including: Morocco, Algiers, Jamaica, Iran, and Chandigarh. Tyrwhitt also made preliminary arrangements while at the congress for two houses to be built at the exhibition by an Indian architect working with Le Corbusier in Paris, who was returning to India to supervise his projects in Ahmedabad and Chandigarh. Tyrwhitt had earlier made plans for three houses to be built by Fry and Drew's team of architects at Chandigarh. [T.56] In this way she facilitated the design and dissemination of prototypes for low-cost housing suitable for Indian conditions that combined a modernist aesthetic with local building materials and methods. She also sought to engage diverse CIAM voices to help shape the discourse at the UN seminar, and recommended that invitations be sent to several CIAM members, based in Japan, Ceylon (Sri Lanka), and Vietnam, a country not officially recognized by the UN.

Upon her return to India in August Tyrwhitt mainly focused on designing the Village Center in cooperation with the Delhi offices of the specialized agencies and, since the exhibit was co-sponsored by the Community Projects Administration, the corresponding Departments of the Government of India: Education, Labor, Health, and Agriculture. A young architect was assigned to design the Center who was "willing and eager to build in rammed earth." Tyrwhitt had to have ... buildings similar to those that can actually be erected by the villagers themselves." [SG.55]

Now optimistic about the project's success, Tyrwhitt began to enjoy life in India. The purchase of a second-hand car made "a fantastic difference," but it was the arrival of her sister-in-law, Delia, in October that transformed Tyrwhitt's experience of Indian life. "She expects a certain amount of comfort and pleasure from her daily life, and this makes me get things and do things I would not have bothered about for myself alone." They moved into two large verandah-lined rooms in a "refined barracks" built for British officers on leave, which they made almost "homelike" with the help of a man who cooked and cleaned as well as took care of the car. "The weather, the surroundings and the people nearest to me are all now congenial." [SG.56, 57] As her private life became more comfortable she became more receptive to the sights and sounds that had initially seemed overwhelming. In Tyrwhitt's circular letters to friends she enthused about "Indian dancing and music and painting ... . Each musician and each dancer is a composer rather than merely a performer: a composer of variations upon a known theme." [T.57] Her letters to Giedion indicate how their collaboration had colored her perceptions, describing subtle cross-cultural influences, such as were evident in a village circus, and signs of continuity and change, such as indications of "the mother goddess ... everywhere, though camouflaged by the Hindu Aryan pantheon of male gods" [SG.57]

Tyrwhitt was saddened to hear that Giedion was unhappy about the English translation of his book on Gropius and wished she could help him. But her work for the UN demanded her full attention, and she still had not edited the CIAM 9 reports; she hoped "Sert will not be angry as his good will may be very valuable to us." She agreed, however, to continue to translate Giedion's two-part article for *Architectural Record*: "The State of Contemporary Architecture"—in which *they* heralded both "a new regional approach" and the emergence of "a new hybrid development—a cross between Eastern and Western" (1954a: 135). [SG.59, 60]

The Village Center

Tyrwhitt intended her working model of an Indian Village Center—a CIAM-inspired Core set amid experimental houses—to demonstrate the necessity and benefits of integrating rural housing policy into the political and economic revival of village life, based on "the restoration of responsibility to the village *panchayat*—a restoration of the self-reliance and pride that made the Indian village of earlier times the real home of thought and culture in India." [T.58] As she brought the Village Center to life, Tyrwhitt must have reflected on Geddes's work in India, as well as Dartington, which, as noted previously, was modeled on Tagore's Institute of Rural Reconstruction at Sriniketan. A direct source of inspiration was Gandhi, whose Basic Schools adopted Tagore's scheme of teaching through crafts. In

December Tyrwhitt visited Sevagram—which Gandhi intended to become his model village—to decide which ideas she could incorporate. In addition to a replica of Gandhi's hut, Tyrwhitt adopted several basic schools-related features, for example, setting the window sills in the schools at 18" above the ground so light could shine directly onto the children's work. She also incorporated a moveable trench latrine, a simple design for a cowshed with moveable mangers, a smokeless chulha (stove), and a simple way of providing light and ventilation at ground level for washing pots in a kitchen. The buildings were constructed of sun-dried bricks or rammed earth, and roofed with thatch or local tiles. There were also two wells and a tank for collecting methane gas from cow dung to be used as fuel for cooking and lighting. [SG.61, 62]

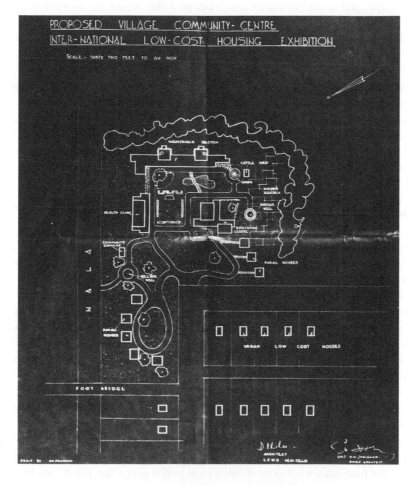

Figure 12.3 Tyrwhitt was inspired by both the CIAM Core and Gandhi's model village, Sevagram, in her design for the Village Center

Tyrwhitt summed up her vision: "The integration of mind and body, hands and the good earth is shown by the careful siting and design of a multiple purpose basic school building," and classes of adults and children were in full swing there; "a small health clinic planned in relation to environmental sanitation needs," was attended by a village nurse; "a crafts center, where sheds were manned by village craftsmen, and in which production was centered on housing; and a seed store and manure producing plant, linked to the cultivation of a vegetable garden which, by being itself linked to the basic school, restarts the cycle of life." [T.59] The simple buildings enclosed an open space with a low platform, the chabutra, or traditional village festival stage, where village performers told stories, sang songs, danced, and did acrobatics. The Village Center expressed the new social imagination that Giedion—and Tyrwhitt—championed as the new regional approach, "that satisfies both cosmic and terrestrial conditions" (1954a: 137).

The International Exhibition of Low Cost Housing and UN Regional Seminar on Housing and Community Improvement

Eleanor Hinder, head of UNTAA's Office for Asia and the Far East Program Division, wrote Tyrwhitt just before the Exhibition and Seminar began: "We had no doubt that the project would be effective under your guidance and feel that its success will be due in large measure to your energy and leadership." [T.60] Their confidence in Tyrwhitt was justified. The President of India opened the International Exhibition of Low Cost Housing on January 20, 1954, the day before the Seminar began in a hall located at one end of the exhibition site. Prime Minister Nehru attended the opening of the IFHTP Southeast Regional Conference, on February 1, 1954. Both took place in the UN Seminar Hall. [T.52] Tyrwhitt ensured that this exhibit included the Moroccan and Algerian grids from CIAM 9, as it proved impossible to arrange a larger showing of material from the congress. Tyrwhitt told Giedion, "I may complain [about the amount of logistical as well as substantive] work but there is a certain excitement about these days, that may keep me awake but is not unpleasant!" [SG.63]

In advance of the seminar, 70 working papers were received and circulated to participants. Official delegations were sent from eleven countries within the ECAFE region, and observers came from four more. India sent representatives of twenty-two institutions closely concerned with housing and community improvement. Five ex-students of the School of Planning came from all over India, and one arrived from Singapore. The American delegation nearly did not arrive at all, however, due to delays in their obtaining "clearance" to represent the US abroad, a consequence of McCarthyism. [SG.64]

Of all the people she met at the symposium, Doxiadis particularly impressed her; she thought he "has a kind of CIAM spirit." Her estimation of Doxiadis as "about the best man we had here" was high praise given the illustrious line up. [SG.64] His paper was on low cost housing suitable for rural areas that could be erected by self-help programs.

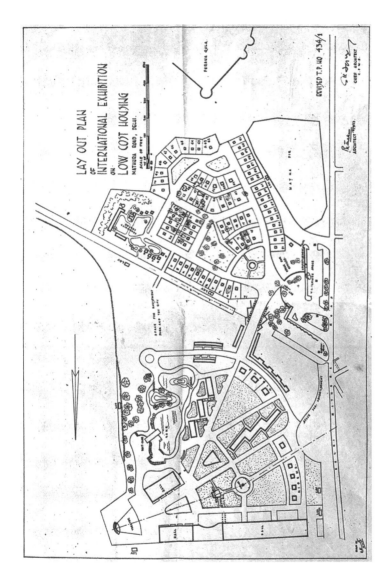

Figure 12.4   The Village Center and model homes were the UN's contribution to India's International Exhibition of Low-Cost Housing. None of the model homes cost more than 5,000 rupees, the equivalent of a little more than 1,000 dollars

Aside from the formal discussions, delegates visited villages built for refugees near Delhi and took weekend tours to Chandigarh and the Bhakra Nagel hydroelectric project, the Taj Mahal, and nearby Fatephur Sikri, a sixteenth-century Mughal city. Le Corbusier made a presentation at Chandigarh, and Tyrwhitt organized a follow-up symposium for a more in-depth discussion of the new capital city. This was a rare opportunity for an international gathering of experts collectively to study path breaking projects in the developing world *in the field,* projects that integrated regional planning, urban design, community development and resource management; those experiences made a deep impression on Tyrwhitt.

Weissmann affirmed in his closing remarks: "The intensity of the urban housing crisis in Asia, coupled with the growing magnitude of the rural problem, suggests the need for a bolder and more imaginative approach. ... The Seminar has shown that the less developed countries are evolving new and more rational approaches and methods of their own more suited to their circumstances and needs." [T.61]

ECAFE met immediately following the UN seminar to translate its conclusions into recommendations to governments in the region. Tyrwhitt and Weissmann worked with ECAFE staff to see which of them could shape UN housing and planning policies during the next two years. Notably among those policies endorsed was a proposal to establish a planning school in Southeast Asia. As her duties in India began to wind down, Tyrwhitt was glad "this period of exile is coming to an end;" she also told Giedion: "I know I have gained a lot in myself from this experience ... the getting up of the Centre *has* been worth while." [SG.65]

Figure 12.5 **Mr. Jawaharlal Nehru, Prime Minister of India with Tyrwhitt, visiting the Village Center**

Chapter 13
Inter-relations:
Communications, Culture, and Urbanism

Global Village

During the previous year, in addition to the trips she made between North America, Europe and India, in preparation for the UN seminar, Tyrwhitt had visited Israel and Karachi, and had made a whirlwind tour of Bangkok, Rangoon, Djakarta, Bandung, Singapore, as well as cities across India. In each place she met with officials, practitioners, and academics concerned with a wide range of issues affecting housing and community planning. In March 1954, at the conclusion of the seminar, Jaqueline and Delia enjoyed a few days of pure tourism. They stopped at Jaipur, a city planned in the early 18th century according to ancient Hindu principles, and also visited the archeological ruins at Mohenjo-daro, one of the world's earliest urban settlements. Inspired by her work with Giedion, Tyrwhitt viewed those historic sites with a keen interest in "evidence of continuity" of ancient culture in modern society. From Mohenjo-daro she traveled alone to rendezvous with Giedion in Baghdad; she enjoyed a four-week "working vacation" with him there and in Egypt. Afterwards, as Giedion puzzled over "the space-conception of the Egyptians and the Sumerians," he wrote Tyrwhitt: "Would be interested what You think about it." [SG.66] Few people, let alone women, could claim first-hand knowledge of so many dimensions of the field of human settlements spanning such a spectrum of space and time.

By virtue of her innumerable contacts and her institutional ties to transnational organizations, such as the UN, CIAM, and IFHTP, and the School of Planning, she was uniquely positioned to forge new linkages and serve as an agent for cross-cultural exchange and cross-fertilization of ideas at a time when technical assistance and international cooperation was still in its infancy—and its future uncertain (Collapse 1954).

Tyrwhitt was eager to expand cross-cultural learning processes in the training of planners and architects. While still in India, Tyrwhitt accepted an invitation to give a two-week workshop in May 1954 based on her experiences for architecture students at Princeton University. For her "Seminar on Village Buildings for the Central Areas of India," Tyrwhitt had the students develop "designs for a prototype village centre for about 3,000 people as part of a village improvement scheme in areas likely to be affected by a new irrigation system of canals connected to a major hydroelectric project." The program called for a primary school, health clinic, and bazaar. [T.61] Tyrwhitt arranged for the students' work to be sent to the head of the Rural Housing Section of the Federal Ministry, an example of her continuing efforts to link her pedagogy and planning practice.

Digging a Hole in the Sand

This was heady stuff to be involved with but at age 49, Tyrwhitt found herself, again, unemployed. She had accepted the UN position in India in part because she needed a job. In April 1953, since funding for the town planning program at the University of Toronto had not been secured, the President would make no binding commitments; he was willing, however, to entertain a three-year appointment for Tyrwhitt as an assistant professor starting in fall 1954, but at a salary she felt was too low. [T.62] Regardless, she went to India assuming that she would return to Toronto and begin her three-year appointment there. At that time she wrote: "There's no doubt the Indian job will be interesting—but my heart was set on the work next year in Toronto." [T.63].

In April 1954, though, Tyrwhitt learned that her position was eliminated. [T.64] The University had cut all funding for planning and transferred all planning instruction to the Architecture School, emphasizing undergraduate courses for civil engineers and architects—in other words, the opposite approach to the Shuster Committee's recommendations in Britain. Nevertheless, Tyrwhitt knew that the Ford Foundation had awarded a two-year grant for McLuhan's study of "changing patterns of language and behaviour and the new media of communication," in which she was named as a member of the faculty team; she received this news in Geneva in July 1953.

Upon learning of the Ford grant Tyrwhitt had written to McLuhan and proposed that they begin by establishing "a common vocabulary among the members of our group," and then conduct a study along the lines of Giedion's "inter–faculty study of methodology." That would suit her personal agenda, since the group might then find it "worth while trying to get Giedion over to frame the programme." But she asked: "Where do I fit in?" Tyrwhitt revealingly compared her work in India to her first year in Toronto:

> One goes on, hopefully, but can't see how anything is working out. ... one seems to be standing still, or to be digging a hole on the seashore which next day has completely disappeared – though it looked like a real hole when you left it ... . [A]fter a bit I shall learn the limit of the tide and dig my hole in the right place. [T.65]

After learning that her faculty position had been cut, Tyrwhitt was determined to dig yet another hole in the sand. She went to Toronto, briefly in April 1954, to try to salvage the situation. The University's press release announcing this prestigious award had touted Tyrwhitt's particular proficiency "at planning co-operation between academic departments"—a skill not attributed to any other team member. [F.2] But the grant did not include any salary-related funding for her. With McLuhan's support, she convinced the University President to permit the reallocation of grant funds to allow her to "carry on under the Ford Foundation," with an appointment as a Research Associate in the School of Architecture. [T.66] This project became her lifeline to an academic position in Toronto.

The Explorations Group

In the interim McLuhan and his main collaborator, anthropologist Edmund Carpenter (1922–2011) had obtained permission to use some of their grant funds to publish a journal, *Explorations: Studies in Culture and Communications*. Carpenter was editor in chief, and the other members of the faculty team were associate editors. Six issues were planned. One issue would be dedicated to Giedion, and "changing concepts of space and time would be a central concern." McLuhan asked Tyrwhitt to write something based on her work in India. [T.67] Instead she gave McLuhan and Carpenter permission to publish the article she had written for the now defunct *Trans/Formation*; a version appeared as "Ideal Cities and City Ideal" in *Explorations 2* in April 1954.

After an auspicious beginning, though, the need for someone able to facilitate collaboration became apparent. In December 1953, McLuhan wrote Tyrwhitt that his "team" was pulling in different directions and the magazine was to provide the structure and impetus for holding them together. He also hoped that Giedion's work would serve as core readings that would provide a focus for the increasingly fractious group. [T.68]

When Tyrwhitt accepted the Research Associate position on McLuhan's team in April 1954 she felt she had to bring that job "into shape or else very publicly get out of it." Sert, who was now ensconced as Dean of the Graduate School of Design and Chair of the Department of Architecture at Harvard, had told Tyrwhitt she would be invited to give a seminar or some lectures in the fall. Tyrwhitt resolved to make her "headquarters next winter in Toronto and travel to and fro a bit from Harvard;" doing so would enhance her ability to work with Giedion, who would be teaching at Harvard in the fall term. [SG.67]

She proposed to McLuhan that she begin in September 1954, "and concentrate on … editing the work that had already been collected so that during this winter we do produce something that can purport to be an 'inter-faculty' study." She added: "I also badly want to get some of my own ideas down, but I am sure the most important thing is to produce the results of some group work." In a pointed reference to *Explorations*, she advised: "We must produce something … that is not just an interesting collection of essays, but does obviously achieve some sort of synthesis and state some sort of theory. How another *Explorations* around Giedion fits into this I am not clear." Since she would be seeing Giedion that summer, she offered to use some of this time to develop aspects of his contribution to their faculty team's work. [T.66, 69]

Tyrwhitt's offer to mediate Giedion's participation in what came to be called the Explorations Group helped convince them to meet regularly over the summer of 1954 to discuss *STA*. "We had decided that Giedion would provide an ideal approach to visual communication problems," McLuhan recalled. "Miss Tyrwhitt ... was of the greatest help here." [F.3] But Tyrwhitt wrote Giedion: "This is all very well, but its still not 'my' field and during the summer (with you I hope) I must draw up an outline of what course I really want to pursue myself." [SG.68] She frankly explained her ambivalence:

> By far my deepest interest ... is to help you get out the 'Continuity" book. Apart from that I have ... (three or four) aspects of 'planning' that I feel very strong about ... but I know that if I really want to 'put them across' I must get down to writing ... . Apart from these two dominating interests, both more or less concealed, there is the need to earn one's living, and determination to do this without having all time eaten up by the job. ... On the other hand Sert holds out a suggestion of something very attractive – and quite up my alley – but I don't know if he can really pull it off or if he really wants me. [SG.69]

Sert had in mind a plan that paralleled the interdisciplinary premise of the Explorations Group: to integrate the three disciplines taught at GSD, architecture, city and regional planning, and landscape architecture, within the framework of urban design (Alofsin 2002, Mumford 2009). Like McLuhan, Sert appreciated Tyrwhitt's proficiency "at planning co-operation between academic departments," but in his first year at Harvard he had many issues to deal with, including upgrading the faculty, healing rifts among the three departments, and revising the curriculum (McAtee 2008). In the summer of 1954 Sert needed Tyrwhitt's help more with CIAM.

Habitat: Geddes, Buber, and CIAM urbanism

Sert had invited Tyrwhitt to give a lecture at GSD in May 1954, concurrent with his announcement of three new urban design courses, beginning with Giedion's seminar on the history of urban design. Tyrwhitt was at Princeton then but went to Cambridge the following weekend to go over the CIAM 9 material, which Sert felt could be the basis of a book on the human habitat. They also discussed strategy for the upcoming CIAM council meeting in Paris, where decisions would be made about the next congress, which was to be devoted to writing a *Charte de l'Habitat*. Both Sert and Tyrwhitt wanted to hold CIAM 10 in September 1955 in Algiers as a small working conference. [SG.70]

The CIAM Council agreed in June that a committee of "younger" members would plan the program for CIAM 10, coordinating with an advisory group including Sert, Giedion, Le Corbusier, Gropius and Tyrwhitt. Two members of the planning committee—later called Team 10— Jaap Bakema and Peter Smithson, had been among a group who had met the previous January at Doorn, in the Netherlands, to discuss their vision of CIAM's future direction. Finding the Athens Charter inadequate, they called for an "ecological" approach to planning and urban design (Mumford 2000: 239–241). The Doorn meeting issued a "Statement on Habitat" declaring: "To comprehend the pattern of human associations we must consider every community in its particular environment."[1] Smithson sketched a diagram based on Geddes's Valley Section to show the relationship of communities—from single house to large city—in relationship to their environments (habitat). Tyrwhitt and Giedion welcomed the Doorn "manifesto" as it supported further development of the concept of the core discussed at CIAM 8, and the synthesis of Geddessian principles and modernist ideals that Tyrwhitt had been promulgating for a decade. [SG.71] Tyrwhitt met with Smithson and others in London in July 1954 to review their draft program. At that time she probably shared her understanding of Geddes's Valley Section, if not her 1951 article. She left that meeting "highly impressed with the seriousness and imagination" of plans for CIAM 10. [L.25]

As previously noted, Tyrwhitt's compilations and careful editing of Geddes's writings had helped spur the revival of interest in his ideas in the early 1950s, especially in Britain. That summer, Lund Humphries asked Tyrwhitt to review a manuscript for a new biography of Geddes. Tyrwhitt—often a severe critic—told Giedion: "It's not good enough—but who is to write the real thing? Not me. Unless of course I could land a good grant … which would enable me … to embark on other ventures of my/our own." [SG.72]

Tyrwhitt's reference to "my/our" ventures signaled the extent to which she thought their work had merged. Giedion asked Tyrwhitt explicitly to blend her words and thoughts on Geddes with his ideas for his first lecture in the Harvard Urban Design seminar. "I would like very much to add some lines on Geddes and his 'Cities in Evolution,' and how he was the first to proclaim a dynamic planning, as you outlined in the introduction and also as it is done in the summary, p. 154 ff." Giedion assumed that Tyrwhitt would understand what to do with text she was already translating, about the space conception of Egypt and Sumer; they "should be in the middle part of the first lecture." [SG.73]

Giedion depended on Tyrwhitt; he was nervous about his debut lecture because people often didn't understand his English. He was ecstatic about its reception: "And 'ME' & 'You' or 'I' & Thou' I was so glad I had your typed manuscript. — T H A N K S. … and your voice." [SG.74] Giedion's use of Buber's "I Thou" terminology both indicated the intimacy of his relationship with Tyrwhitt, and signaled their shared identification with Buber's dialogic philosophy.

---

1   See Doorn Manifesto, *Team 10 Online*, available at http://www.team10online.org/ (accessed June 16, 2012).

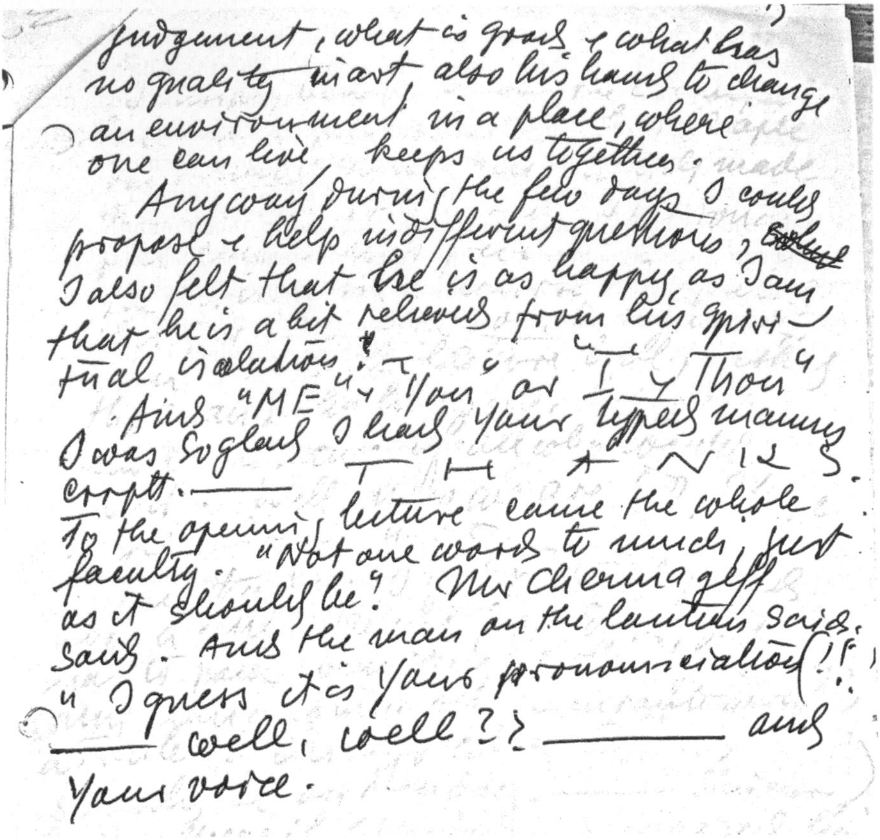

judgement, what is good & what has
no quality in art, also his hand to change
an environment in a place, where
one can live, keeps us together.
Anyway during the few days I could
propose & help in different questions,
I also felt that he is as happy as I am
that he is a bit relieved from his spiritual isolation!
And "ME & You or I & Thou"
I was so glad I had your typed manuscript. —— T H A N K S.
To the opening lecture came the whole
faculty. "Not one word to much, just
as it should be." Mr Chermayeff
said. And the man on the lantern said:
"I guess it is your pronounciation(!)
—— well, well ?? —— and
your voice.

Figure 13.1 Giedion acknowledged privately his dependence on Tyrwhitt

Explorations Group Work

Once she returned to Toronto in August 1954, Tyrwhitt was named secretary of the Culture and Communications Seminar, which was chaired by McLuhan. "I willingly accepted as I shall have some time," she told Giedion, "and it gives me some status that I may need this year." [SG.75] Tyrwhitt found the work of group to be "confusing," but she soon imposed discipline—a characteristic role for which Giedion fondly referred to her as "general"—and instituted the keeping of minutes of their weekly meetings. Tyrwhitt and all but one of the students were new to the group in its second year, so the seminar was practically starting anew.

Tyrwhitt quickly established her key role in the Explorations Group not only as its liaison with Giedion, but also as an interpreter and guide to his writings. At the second meeting of the seminar the discussion was based on her revised version of Giedion's unpublished notes on Anonymous History—the "origins of everyday life"—and extracts she selected from *MTC*. [F.3] McLuhan, Carpenter, and Tyrwhitt drafted a list of topics for papers, which fell into four groups: "Anonymous History and Communication, particularly in the entertainment world; the Artist as discoverer; Mass Media as a means of Escape; and 'unawareness,' particularly with regard to the physical environment." Tyrwhitt was interested in the last topic. [SG.76] Giedion would critique the papers when he visited the seminar in January 1955.

The Moving Eye

In November, Tyrwhitt wrote a paper for the seminar based on her experience in India; "Fatehpur Sikri: The Space Concept of a Moghul City Core," was published as "The Moving Eye" in *Explorations 4* (1955). "The Moving Eye" had a polemical tone, appropriate for the limited circulation *Explorations*. Tyrwhitt began by musing on her visit to Fatehpur Sikri, a city planned and built in the late sixteenth century in Uttar Pradesh, to serve as the capital of the Mughal Empire. Abandoned by the Mughals after a few decades, Fatehpur Sikri was archeologically recovered in 1892 and preserved as an example of the excellence of Mughal architectural ensembles. Tyrwhitt wrote that upon entering the ceremonial core of the silent city—the *Mahal-i-Khas*—"one experiences a rare sensation of freedom and repose" (1955a: 90–91) The visitor "becomes an intimate part of the scene, which does not impose itself upon him, but discloses itself gradually to him, at his own pace." Struck by the sophistication of the asymmetrical spatial composition of the *Mahal-i-Khas*, Tyrwhitt's essay purported to be an attempt to find its organizing principle. But her main point is that the order underlying the *Mahal-i-Khas* was not evident from any single perspective. The pleasures of the place—the felt pattern of coherence—were revealed to the "moving eye." She thought the key to this composition was not found in Western art.

Tyrwhitt noted that the "moving eye" corresponded to a technique developed by Classical Chinese painters of parallel perspective with no vanishing point; the spectator's viewpoint shifts through a scene. A similar pictorial language—free "of the restraining blinkers of the single viewpoint"—existed in early (pre-scientific) Western civilization.

She also referred to a thesis she had *not* read by Doxiadis, who attempted to explain the principles underlying the composition of the Acropolis at Athens (1955a: 93–94). (Her understanding of Doxiadis's thesis was based on a conversation they had in India, possibly at Fatehpur Sikri.) Doxiadis used a geometric system based on a triangular field of vision to demonstrate how a spectator would view the Acropolis from a series of vantage/vanishing points as a sequence of coherent architectural scenes. Tyrwhitt suggested that this panoramic field of vision is closer to the spatial pattern of the *Mahal-i-Khas* than "a single piercing view that demands a central [that is vertical] point of interest."

Echoing Moholy-Nagy, Kepes, Giedion and many other artists and intellectuals at that time Tyrwhitt concluded (1955a: 94–95): "It is now nearly half a century since western artists and scientists started to break away from the tyranny of the static viewpoint—the conception of a static object and a static universe— to rediscover the importance of vision in motion." Therein was the key to "our contemporary urban planning problem: to find ... an intellectual system that will help us organize buildings, color and movement in space" that incorporated multiple perspectives, and fused artistic and scientific sensibilities.

Tyrwhitt's text perturbed Giedion, who detected in it "drops" of his ideas— probably referring to "the dominance of the vertical," an idea they had been discussing since at least 1949— which he felt were premature for publication. [SG.77] Tyrwhitt deftly deflected his criticism, and published an expanded, illustrated version of this essay (minus that phrase) as "Fatehpur Sikri," in *Architectural Review* in 1958. An anthology edited by McLuhan and Carpenter, *Explorations in communication* (1960) included both Giedion's "Space Conception of Art" and Tyrwhitt's "The Moving Eye."

East West Dialogue

Tyrwhitt asserted in "The Moving Eye" that one exemplar of a holistic perspective could be found in Japan—"before the penetration of Western thought"— where there was no distinction between "fine art" and "'the way of doing' things, whether solving a problem, building a house, or preparing tea" (1955a: 94). Her claim exemplifies the *Japonisme*—the admiration of all things Japanese—which was enjoying a mid-century US revival, part of a larger phenomenon, the creative force of the interaction between the civilizations of the East and West, widely acknowledged as generative in the Renaissance and the emergence of modernism in Europe by World War I (Meech 1988, Lancaster 1983, Sullivan 1989). Japonisme was particularly strong in Tyrwhitt's circle:[2] her aunt Ursula had visited Japan before World War II, Delia worked there during the Occupation, Wells Coates spent his youth there, and at CIAM 8 she met architects Kunio Maekawa (1905–1986) and Kenzo Tange (1913–2005). By the mid-1950s growing interest in Japanese design drew more and more Western architects to Japan "to see for themselves" (Koike and Hamaguchi 1956: 18). Gropius spent four months there in 1954.

Giedion frequently reiterated the comments he wrote in 1953, in his Forward to the Japanese edition of *STA*, which, as revised and translated by Tyrwhitt, spoke of his hope for a slow transformation that would lead to a "cross between western and eastern mentality ... the new myth of our period." [T.70] And, as noted previously, while in India Tyrwhitt translated Giedion's two-part article for *Architectural Record*: "The State of Contemporary Architecture"—in which *they* heralded the emergence of "a new hybrid development—a cross between Eastern and Western" (1954a: 135).

---

2   Thanks to Catharine Nagashima for bringing this to my attention.

The cross-fertilization of Eastern and Western social-aesthetic ideals in an emerging global culture became an important theme in Tyrwhitt's work. The relevance of this line of thought to urban design practice and pedagogy was on her mind in late 1954 as she prepared a lecture to give at Harvard. In addition to a public lecture on the Village Center she gave a talk to the Introductory Survey of Planning Course taught by Isaacs, who suggested she focus on "the human problems of housing and planning in India." [T.71] Tyrwhitt rehearsed her ideas in a talk in Toronto—later published in the *Journal of the Canadian Institute of Architecture*—in which she emphasized the hybrid and experimental nature of the housing being built at Chandigarh:

> a cross between traditional habits and westernized ideas – in terms of brick construction – the most economic and efficient building material available – and the requirements of the Punjab climate. They are deliberately experimental ... . Each design has led to another, and there is no doubt that a new ferment had started in the design of dwellings for India that may be able to birth a new and truly Indian development of domestic architecture. (1955b: 16)

Invitation to Harvard

On January 4, 1955, the day after Tyrwhitt's Harvard lecture, Giedion informed her confidentially ("behind a stainless steel curtain"): "You will be appointed as a assistant member of the [faculty] ... even Isaacs agrees." [SG.78] Tyrwhitt received the news with cautious joy: "I dare not let myself really believe yet that there is really a place for me at Harvard; that the struggling part of 'earning a living' may be over ... and I can really start to work quietly and steadily at a job that's worth while, among people whom I both like and respect. This is something that has seemed ... as though it could never happen to me." She worried, though: "so many things have been withdrawn almost as soon as I'd dared to hope they were true—so why not this." [SG.79, 80]

Tyrwhitt was, indeed, appointed Assistant Professor of City Planning in February, becoming the first woman to hold a full time faculty position at GSD. She was disappointed that the appointment was for one year only, but she was cautiously optimistic: "The rest can come with time and luck." Giedion advised, "The main thing for you will be: to be as *female* and silent as possible, and as little as possible a school master; *especially* in this tense atmosphere!" [SG.81] Giedion's advice sounds sexist today, but was a plausible strategy for Tyrwhitt to adopt given the chauvinist atmosphere within academia.

Toronto: Saving the Town Planning Course

In the interim, Tyrwhitt had been lobbying to restore the graduate planning program at University of Toronto. The collapse of the course had also distressed and bewildered Carver. "It is a long story of misunderstandings and, primarily, the collapse has been due to lack of a strong, knowledgeable and vocal leadership," he told Tyrwhitt in May 1954; Carver had been unable to supply such leadership from outside the university. However, he hadn't given up: "I intend to go on bothering people until the University has set up a proper course, and even dare to hope that this might be done in 1955. But it seems that this will require much conspiracy and effort—and where are the conspirators?" [T.72] Tyrwhitt became a co-conspirator, and, as Carver recalled: "her tempestuous energy swept across the campus" (Carver 1975: 120).

Typically, Tyrwhitt felt responsible: "IF I'd NOT gone to India I guess I could have prevented the destruction of the course from being on quite such a wholesale line … . On the other hand maybe they really did just want to get rid of me personally—*reculer pour mieux sauter* with someone else in charge." [T.73] Tyrwhitt took action on many fronts: speaking to local planning groups; working with the senior architecture students on the planning aspects of their design studio (the redevelopment of Toronto Island); and organizing an extension course for engineering students (which she taught in the spring term). Giedion cautioned Tyrwhitt about helping others while sacrificing her health:

> It is fine that you organize the planning course for Toronto. But don't forget, what you forgot last time, to take care, from teeth to toe, inclusive hair & hands. It is at least as important! [SG.82]

(However Giedion had his own agenda. "I would like that you are not completely absorbed by others. The things WE are working on are—for me at any rate—so exciting because the ideas are chained together in a new & loose way & yet, I hope also, have a common blood stream going through them.") [SG.83]

Tyrwhitt's initiatives provided the added impetus that enabled Carver and others to consolidate support for reviving the planning program. In November, Tyrwhitt boasted to Giedion: "The result of my work here seems that we shall get an endowment for a chair of planning (not yet certain but hopeful), and also the prestige of the school vis-à-vis the town and the university has gone up as the students did a magnificent job on a new plan for Toronto Island." The endowment— for the first chair of town and regional planning in Canada—was settled by March. Tyrwhitt then intended to draw up the curriculum, leaving enough flexibility "so that the new man can adjust it to his liking. But there must be something already set as the co-operation of a number of faculties is needed and I want to make sure this is secured before I go." Tyrwhitt persuaded the university to hire Gordon Stephenson "for my job here." Tyrwhitt felt her colleagues were "relieved [she] was not in the running." [SG. 84–87]. Carpenter opined: "Jackie Tyrwhitt knew how to translate thought into reality. Never thanked, never credited, she helped change Toronto" (2000).

Before leaving Toronto, Tyrwhitt orchestrated what McLuhan considered the Exploration Group's "major effort:" a test of visual (un)awareness carried out with 800 students at Ryerson Institute. This experiment was the result of Tyrwhitt's determination, noted previously, that the group produce something that achieves some sort of synthesis and states some sort of theory. Students were asked in December 1954 to complete a questionnaire "to discover what particular aspects of the environment of the campus communicated to them—or in other words— registered in their minds." Tyrwhitt and Williamson reported on the results in "The City Unseen," published in *Explorations 5* (1955). Tyrwhitt told Giedion: "I think there are a few new and useful ideas in it." She brought the raw material with her to Harvard, "so that it can be re-used if worth while. But it's more in Kepes' line." [SG.88] Kepes was collaborating with his former student, Kevin Lynch, in a Rockefeller Foundation-funded project at that time that was utilizing a similar approach to studying perception of the urban environment.

Part V
1955–1972

Chapter 14

Urban Planning and Design Education
1955–1960

Tyrwhitt turned 50 in May 1955, and entered a new, highly productive phase of her career. As Assistant Professor of City Planning at Harvard, she helped Sert to introduce a new urban design curriculum at GSD, which involved team-teaching cross-disciplinary classes. She introduced her Geddessian line of modern planning thought into their content as well as into broader discussions on urban design sponsored by the school. She also helped produce illustrated publications that translated university-based discourse about the forces shaping the urban environment for a general readership. Tyrwhitt played a major role in a partnership between Harvard and the UN to establish a new school of planning in Indonesia, thereby facilitating the internationalization of the university and planning education. In a parallel effort on behalf of Doxiadis, from her base at Harvard she launched an information service to support his staff and the growing number of consultants working in the developing world, a domain of practice spurred by the growing role of the UN in the field of human settlements. And she continued to devote as much of her time as possible to Giedion's publications. One theme informing all of those undertakings was the creative dialogue between and cross-fertilization of Eastern and Western social-aesthetic ideals.

Preparing for Harvard

Sert had recruited Tyrwhitt as a key ally to help him realize his vision for a new program in urban design as "the meeting ground of architects, landscape architects, and city planners." In addition to knowing Giedion well, Tyrwhitt knew many members of Sert's predominantly new faculty, most of who had alliances to CIAM and/or Gropius. She understood that as a woman and a non-traditional academic she would have to operate from a marginal position. And she recognized her influential role—the outsider as insider—early on: "Sert … makes me feel that he relies a good deal on my ability to act as a go-between in the departments—and someone to whom he can talk freely and who will let him know of problems before they materialize." [SG.89]

She initially reported to Isaacs, whom Sert had appointed in 1953 to serve as chair of a newly merged Department of City Planning and Landscape Architecture. This short-lived departmental merger—it ended in July 1956—reflected Gropius's belief that planning education should be related to physical design (and that landscape architecture had become indistinguishable from physical planning); it also underscored Sert's decision to maximize the limited financial resources of the school by "teaching physical planning only and getting outside help to cover all other related subjects." (McAtee 2008: 176–177). While Tyrwhitt did not share Sert's and Gropius's architectural biases, she agreed that "without a structurally creative imagination even the best collection, analysis and synthesis of place, folk and work could not result in a worthwhile habitat for man"—a perspective she credited to her association with CIAM. [E8].

Sert assigned Tyrwhitt the pivotal role of co-teaching, with Serge Chermayeff, the Environmental Design Studio—a full year course all entering GSD students were required to take—and to redesign the course. "The opportunity is a terrific one: it's the really vital year—that sets the standard for everything ... . And one is working with the whole crowd—architects plus landscape plus planners. If one can't get ideas of integrated perception and operation over here ... well it's a magnificent challenge." [SG.90] Additionally Isaacs invited Tyrwhitt to improve the lecture course that she would teach: Introductory Survey of Planning. [T.74]

It was hard to find the right balance between planning and design in the studio. "So far Serge ... is willing to go quite a reasonable way with me on the planning side," she told Giedion. Isaacs, however, was reluctant to relinquish control over the lectures. Tyrwhitt picked her battles: "I told Sert that I didn't think one could get everything straight this year without more trouble than he wanted, and he will be happy if the Environmental Design studio course ... is kept at a (is raised to a) good level." [SG.91, 89]

Architect Eduard Sekler, who had studied at SPRRD in 1946, and whose career Tyrwhitt had nurtured by helping him find work in the US, was a visiting professor at Harvard that year; she coordinated their courses for first year students, and gave several lectures in his class on architectural theory, including one on Fatehpuhr Sikri. Tyrwhitt also began to play a largely hidden role facilitating collaboration between Sekler (who became a regular member of the faculty in 1956) and Giedion in co-teaching the new four-semester survey of architectural history and urban design seminars. [SG.92, 93]

CIAM Swan Song

In June 1955, in addition to planning her courses for Harvard Tyrwhitt also prepared for a CIAM Council meeting to be held in Paris in July 1955. Plans for CIAM 10—which she felt "must be the last meeting of the 'old guard'"—remained vague [SG.91]. The Team 10 committee was split into factions; its effort to specify a program of work for the congress—ostensibly to prepare a Charte d' Habitat— had been riddled by conflict (Mumford 2000: 241–244). Misunderstandings were compounded by cultural sensitivities associated with translating between English and French. Le Corbusier proposed an endgame: postpone the congress for a year to allow Team 10 to finalize its program; meanwhile, CIAM members would meet as planned in September and draw up a *Charte du Logis* (statement on housing rights) that would sum up the experience of CIAM.

Tyrwhitt considered Le Corbusier's proposal "a masterpiece of 'finangling' [sic]: the invention suddenly of the *Charte du Logis* was very Corbu! Of course he has it all written down already." She urged Giedion not to be upset by the situation:

> The CIAM situation is of course confused, but we can at least hope ... that a group will be set up ... to produce this Charte du Logis—which is of course the swan song of the CIAM of the last 25 years. ... Maybe the younger crowd will, after all, show themselves able to continue with CIAM and the intention to produce a Charte de l'Habitat—or something else. ... Presumably the Charte du Logis, when framed, will have to be presented to some Congress or other ... Perhaps this will be the Final Occasion? [SG.92]

Before they went to Paris, Tyrwhitt helped Sert organize a meeting of potential North American CIAM groups from NY, Boston, Toronto, and Philadelphia, to discuss the work done at CIAM 9, which Tyrwhitt was bringing to Paris, and the work they planned to show at CIAM 10. Tyrwhitt was hoping to energize younger members in North America, who did not share the rancor that was roiling some members of Team 10 in Europe—who aimed their invective personally at Tyrwhitt as well as Sert and Giedion: the "Harvard professors." [SG.91]

At the Council meeting in Paris, Le Corbusier's proposal was accepted. CIAM members met in September at La Serraz, Switzerland, to work on a *Charte du Logis* and determine the program for CIAM 10, which would be held in Dubrovnik, in August 1956, due to the revolution underway in Algeria. Tyrwhitt was charged to work with Sert, Gropius, and Giedion to prepare the final English text of the document produced at La Serraz, an outline of the Charte, titled: "The Dwelling: Statement of Principles." As proposed by Giedion, the theme for CIAM 10 was now, "The Habitat: Problems of Inter-Relationships." In terms that again called to mind Buber's philosophy, Giedion stated: "The city is above all a matter of interrelations, encounters, the confrontation of you and me, the inter-action of the habitat of the individual with that of society." [L.26]

Tyrwhitt then helped Giedion underscore the significance of "the confrontation between you and me" when she produced *his* swan song for CIAM: *Architecture You and Me: The Diary of a Development* (1958) an expanded version of *Architecktur and Gemeinschaft* (1956)—a collection of his articles and lectures written between 1937 and 1955. In his "Finale," written in 1955, Giedion affirmed: "The four main functions of urban planning: living, working, recreation, and communication, as they were stated by CIAM in 1933, have lost their balance and their inter-relationship ... . The demand for the reestablishment of the relation between 'you' and 'me' leads to radical changes in the structure of the city" (1958: 203). Giedion acknowledged Tyrwhitt's substantial contribution to this book by agreeing to share a percentage of its revenues with her; characteristically he was less generous about publicly sharing credit. [T.75] In the last sentence of his Foreword Giedion stated simply that Tyrwhitt did "the revision of English texts and translation of material from German or French" (1958: vi).

The First Urban Design Conference

An important part of Tyrwhitt's job in the fall term of her first year at Harvard, was to help organize a major conference on urban design scheduled for April 1956. Sert convened the event "in view of the great interest in urban renewal and urban redevelopment and the continued growth of cities in this country" (1956: 492). A major incentive was the availability of federal funding under the Housing Acts of 1949 and 1954. Federal and state legislation made financial support for urban redevelopment projects contingent on the preparation of plans for land use, open space, neighborhood facilities, and infrastructure improvements. There were already many city-wide and metropolitan redevelopment programs underway or about to be launched—and a shortage of trained planners to do the work (Feiss 1954). The premise of Sert's conference—and new urban design curriculum— was that this type of work called for a more comprehensive approach to physical planning. By focusing on aesthetic aspects of urban design, Sert hoped this conference, the first of a series, would build consensus among faculty of the three GSD disciplines on the "common basis for joint work" (Planners 1956)—and thus avoid academic turf battles. Tyrwhitt was part of a planning committee, but in fact she was the principle organizer of the conference and the designated rapporteur—a now familiar behind the scenes role.

Although Tyrwhitt was not among the speakers Sert echoed in his introduction Tyrwhitt's words at CIAM 8, on the concept of the urban constellation: "every American city, because of its growth has to break up into constellations of communities. The necessary process is not one of decentralization, but one of re-centralization." [L.27] This reference would not have been lost on the many people in the audience that had ties to CIAM. Over 200 leading planners, architects, and landscape architects attended.

As rapporteur, Tyrwhitt framed the major themes of the urban design conference—UD 1—as they were disseminated to stimulate and inform broader public discourse. She assembled and selected the excerpts of speeches and discussions that were published as a feature article in the journal *Progressive Architecture* in August 1956. The conference was newsworthy in part because it signaled the emergence of a reaction—voiced most famously by critic Jane Jacobs (1916–2006)— to the anti-urban sentiment prevalent in American discourse (Mumford 2009: 122). Jacobs, the only female panelist, was a last minute substitute for her boss, editor Douglass Haskell, who asked Sert's permission to include another woman besides "Miss Tyrwhitt" (Laurence 2007: 10).

CIAM X: Charte d' Habitat

After the conference, Tyrwhitt and Sert got down to work on the English version of the *Statut du Logis* for the forthcoming CIAM 10. She compiled a draft for Sert, Gropius and Giedion to review. "I know this Logis business is not the things that deeply interests you (or me)," she wrote Giedion, "but there it is—it is the 25 years of CIAM to some extent, and it must be set down or we shall be back in the morass at Dubrovnik, instead of moving to your theme—the structure of the new urban setting—which would really be something worth discussing." [SG.96]

Giedion was convinced that the tenth congress had to prepare a *Charte d'le Habitat*. Tyrwhitt had tried to persuade him to stick with Le Corbusier's proposal to leave this task to the younger generation. For this reason Le Corbusier did not attend the Congress. "To re-introduce the *Charte* idea now would make utter confusion," she argued. [SG.95] But Giedion prevailed and even wrote a draft that served as the basis for discussion at the congress by a commission that included Tyrwhitt, Sert and Team 10 member Jerzy Soltan, a Polish architect (who joins the GSD faculty in 1958). The congress resolved that the CIAM leadership would resign at the end of the year and a new regime would be established. It was also decided that Sert would complete the *Charte de l'Habitat* at Harvard, which assumed that Tyrwhitt would activate that process.

Tyrwhitt made the arrangements for a group of CIAM members to travel together from Venice by boat to Dubrovnik, an enjoyable excursion intended to make the final congress memorable. A student who attended recalled: "Tyrwhitt was the *spiritus agens* of this congress too. She was ... the person who was worried about the old Giedion [then 68], ... while simultaneously worrying about the feelings and actions of some of us—the youngest participants" (Music 1985).

Inter-relations: Doxiadis, the UN, and Harvard

After CIAM 10, Tyrwhitt traveled from Dubrovnik to Athens to work with Doxiadis, with whom she had kept in touch since their meeting at the New Delhi seminar in 1954. Tyrwhitt had been very impressed with Doxiadis and the feeling was mutual. In early 1955 she wrote him about her appointment to Harvard, and enclosed a copy of "The Moving Eye," which referenced his doctoral thesis. Doxiadis hoped to work with Tyrwhitt again and suggested they might "meet ... when you will be in Europe." [T.76]

When they next met in London, in August 1955, Doxiadis's engineering consulting firm, Doxiadis Associates (DA), had just been contracted by the Iraq Development Board to prepare a housing program for that new nation, and his firm was growing (Pyla 2008). He asked Tyrwhitt if she would take responsibility for the preparation of a monthly bulletin of international information that would be useful to his staff, particularly those in the field, something similar to the APRR Information Bulletin she had earlier produced. With the end of CIAM in sight, Tyrwhitt would have more time, so she accepted, but with a condition: the bulletin would also be sent to UN experts working in developing countries.

Tropical Housing and Planning Monthly Bulletin

Tyrwhitt produced the first issue, *Tropical Housing & Planning Monthly Bulletin*, in October 1955. The purpose of the new venture, "apart from providing information for C.A. Doxiadis that may be useful to him in his many current tasks in the Near East and in South East Asia, is to build up contacts between those with similar problems who are working in this field," she explained in a letter to potential subscribers in December 1955. [T.77] For the first seven years, Tyrwhitt used an electric typewriter to prepare a master copy of the journal each month, which she sent from her home in Cambridge to Athens for reproduction and distribution. This was the beginning of the journal later known as *Ekistics*, which is the term Doxiadis coined to name "the science of human settlements." Tyrwhitt remained closely associated with the journal until her death.

She began producing the *Bulletin* in an "experimental and exploratory way." [E.9] At first the contents consisted of extracts from and digests of books, articles, UN documents, conference proceedings, and academic reports from various countries. She also included field reports from people with first-hand knowledge of housing and planning in tropical countries. While maintaining a tropical housing focus, Tyrwhitt addressed the range of subjects encompassed by expanding UN programs that gave high priority "to social aspects of housing and community planning and the mobilization of self help," since those aspects were "particularly important for the developing regions of the world" (Weissmann 1978: 230).

With her quasi-insider status at the UN, Tyrwhitt was able to use the *Tropical Housing & Planning Monthly Bulletin* to disseminate UN documents not otherwise easily available, and as a forum for debate over evolving UN policies. She inserted her editorial voice through the selection, length, and tone of her abstracts, which often included her own comments. Tyrwhitt's perspective was evident, for example, in her abstract of a rough draft of a report by Catherine Bauer on "Economic development and urban living conditions" in the February 1956 *Bulletin*. Bauer praised the surveys undertaken by local Planning Research and Action Institutes in Uttar Pradesh, India. Tyrwhitt commented: "This concept of a 'community self-survey' was the key to Patrick Geddes' whole philosophy of regional planning, greatly influenced by his own long Indian experience." Tyrwhitt closed by quoting Mumford's reinterpretation of Geddes's ideal: "The task of regional survey, then, is to educate citizens: to give them the tools of action, to make a ready background for action, and to suggest socially significant tasks to serve as goals for action. Ultimately, this becomes the essential duty of every vital school, every responsible university." [E.10]

In spring 1956 Doxiadis invited Tyrwhitt to work for a few weeks in his Athens office "in order to get your advice, opinion, and criticism on several of our plans." Tyrwhitt spent her time in Athens reviewing DA's seven volume Iraq report. She found it to be "a good job on the whole, and the main points—the Doxiadis points —are very sound and forward looking." Moreover, she enjoyed working with Doxiadis, whom she considered "one of the few people who understands my own profession from the inside, and knows me from that point of view." Doxiadis then offered her "a large salary to work with him in the middle east," a job she briefly considered accepting. However, after suffering a severe asthma attack on the voyage back to the US, she concluded: "it would be crazy to jeopardize in any way my job at Harvard. It's the first time since I left the APRR in 1948 that I've had any sort of security for the next year." [SG.96, 97] Another factor affecting her decision might have been that while in Athens Tyrwhitt had learned of her father's death. This proved to be another turning point in her life.[1]

The Second Urban Design Conference

Tyrwhitt reveled in Harvard's Ivy League ambiance. Her teaching load was substantial, though: 60 students in her Environmental Design Studio course, and 86 in her Introduction to Planning lecture course. Knowing that Tyrwhitt gave every student her full personal attention, and believing that her asthma attacks were due to exhaustion, Giedion warned: "Keep back now with your offerings to other people!" Tyrwhitt agreed: "I am sure [Sert] will not want me to take on other commitments—he will have plenty of his own to keep me busy though not paid!!" Even before the term began, Sert had instructed that all correspondence concerning the second Urban Design Conference (UD 2) be directed to her. [SG.98, 99]

To prepare for UD 2, scheduled for April 1957, Tyrwhitt organized and was the sole woman to take part in a Round Table in late November 1956. [T.78] Based on the success of the first conference, the next task was to operationalize this new field, whether it was called urban design or physical planning. Round Table participants agreed to narrow the field to "the design section of the planning process." Discussion followed on the goals of urban design so delimited, and goal statements were formulated to frame further discussion at the second conference in April. [L.28] Sert hoped that conference would result in a "clear and concise statement about planning philosophy" (City 1957).

The Round Table prompted an interesting exchange between Tyrwhitt and Kevin Lynch, then an associate professor of city planning at MIT. Lynch and Kepes—who were both Round Table participants—were working on a study of perceptions of the urban environment that was similar to the experiment Tyrwhitt had earlier conducted in Toronto. Both studies addressed a problem identified by the Round Table: "In their present shapes, cities are not conducive to a truly urban and urbane way of life as they do not have a comprehensible visual order on the scale of the individual human observer to compensate for the increasing incomprehensibility of their total environment." [L.28] Shortly after the Round Table, Tyrwhitt and Lynch shared their results. Lynch concluded: "the specific

---

1   Personal communication with Catharine Nagashima, June 12, 2012.

results of the Toronto study mesh well with our own investigation, to the extent that the results can be compared." [T.79, 80] Kepes and Lynch presented their work on the perceptual form of the city as featured speakers on the opening panel at the conference, which drew a capacity crowd of about 400 people.

As usual, Tyrwhitt stayed behind the scenes, drafting the resolutions that were voted on, taking notes of the proceedings, and editing the reports. (She also organized an exhibition about urban design in the US since World War II, which made manifest the consensus achieved at the first UD conference.) Once again, she demonstrated her ability to create forums that facilitated meaningful exchanges of information, cross-fertilization of ideas, and generated new syntheses. Sert's remarks at the Second UD Conference again made it clear that he had adopted Tyrwhitt's concept of the urban constellation as his own:

> [N]atural growth ... is growth by cells that form elements or parts. ... The patterns resulting from such growth could produce an urban constellation. I believe that in such a structure an urban and urbane way of life can be developed. But the key to such a way of life lies in the preservation of human contacts, and ... this calls for the breaking-up of these vast regions into ... differentiated units. These units would develop around cores and the process would be one of recentralization as against decentralization. [L.29]

Adept at mobilizing "the collective brain," Tyrwhitt was unconcerned with the "ownership" of ideas.

The Shape of Our Cities: Harvard, CIAM, and the UN

Sert also enlisted Tyrwhitt to work with him on a project funded by the Ford Foundation's Fund for Adult Education, called "The Shape of Our Cities: A Series of Experimental Study-Discussion Programs on Urbanism." The aim of the project was two-fold: to inform participants about the problems of uncontrolled urban growth and alternatives to it; and to develop and test new visual materials and techniques for use by civic groups. Tyrwhitt developed this project in close consultation with Martin Meyerson, who in spring 1957 had been appointed Professor of City Planning and Urban Research, and Director of the Center for Urban Studies, and joined the GSD faculty in the fall. Ten discussion topics were included: the reality of change; the growth and scale of cities; the individual and the family; the family and the community; man and the automobile; community meeting places; the urban region; the residential sector; from utopia to reality (that is, implementation); and the shape of our cities tomorrow (metropolitan governance). Sert wrote brief introductions to each topic; Tyrwhitt selected illustrations and wrote one-page introductions to them. [L.81]

**Figure 14.1   The Shape of Our Cities discussion series (1957) informed the
public about alternatives to the predictable problems caused
by uncontrolled suburban sprawl**

"The Shape of Our Cities" discussion series was tested with several citizens
groups in fall 1957, with mixed results. Ultimately, the awkward format (a set
of folding panels) made reproduction costs too high, and the Fund for Public
Education decided to drop the experiment. However, Tyrwhitt repurposed the
material for both CIAM and the UN.

Tyrwhitt had concluded in January 1957 that "The Shape of Our Cities" was
"really a draft of the *Charte de l'Habitat*." Sert agreed, and wanted to produce
extra copies of the first draft to send to CIAM groups, saying 'here is an American
version of the *Charte*, will you comment on what changes you think need to
be made for your country?'" [SG.100, 101] Around the same time Weissmann

commissioned Tyrwhitt to prepare an exhibit based on "The Shape of Our Cities" materials for display at a UN-TAA-ECAFE Seminar on Regional Planning to be held in Tokyo in summer 1958. Tyrwhitt arranged for subsequent exhibitions in India, Pakistan, Ceylon and Indonesia after Tokyo.

Tyrwhitt worked closely with Weissmann—who was then in charge of the Housing Building and Planning Branch—thanks to Doxiadis, who paid her to spend one day a month at UN headquarters and encouraged her to assist Weissmann. [D.1] Sert was happy about that; he knew it would be useful to have inside information on the UN's expanding program on urbanization in underdeveloped countries. [SG.102] Her work with Weissmann and Doxiadis enhanced her understanding of the universal aspects of "The Shape of Our Cities" material, enabling her to see how it connected with the aims of the UN seminar. The paper Tyrwhitt co-authored with Sert for the UN seminar stated that they hoped their exhibit would

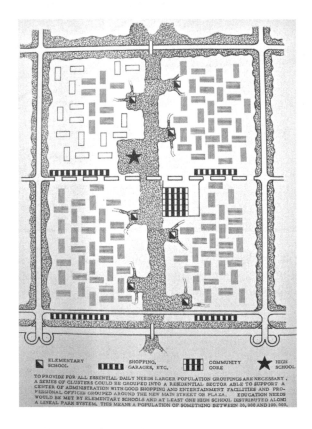

Figure 14.2   The Shape of Our Cities series endorsed Tyrwhitt's concept of the basic "social unit," which, when paired forms an "urban unit" organized around a high school, where "true social integration will become increasingly easy and normal"

"bring to light some aspects of the shaping of man's urban environment which are common to all cultures." [L.23] A universalistic emphasis was also consistent with the Ford Foundation's new international scope: "The world grows smaller and American responsibilities and opportunities larger, dictating the need for greater understanding ... of other nations, of the differences that separate us from them, and of the greater things we do or should hold in common" (Ford 1956: 11).

Doxiadis was very happy with the work Tyrwhitt was doing for him and in March 1957, reiterated his job offer: "Jackeline! [sic] I think you must join us in Athens for a few years and as soon as possible. Please don't commit yourself to anything before we walk [sic] about our plans." [D.2] His firm had just been contracted to produce a national housing plan for Lebanon and to design schools in Pakistan; it was quickly earning a reputation as an organization capable of handling large-scale projects for governments and international organizations (Kyrstis 2006).

Among other things, Tyrwhitt helped Doxiadis stay abreast of the UN's activities, which she tracked in the monthly *Bulletin*. At its May 1957 session, ECOSOC approved a long range program in *community development*—defined as "a process by which the efforts of the people themselves are united with those of governmental authorities, to improve the economic, social and cultural conditions of communities, to integrate these communities into the life of the nation, and to enable them to contribute fully to national progress." [E.11] Based on the experience gained in the previous decade, UN policy in the field of human settlements was evolving, along with the realization of the need for comprehensive programs aimed at as many issues of human settlements as possible. Weissmann (1978: 231) credited Tyrwhitt as being among those that helped formulate this evolving UN policy; the UN's Long-Range International Program of Concerted Action in the Field of Housing and Related Facilities, adopted in 1959, embodied this more comprehensive, top-down/bottom-up approach. UN initiatives to strengthen international action for community-based development resonated with the efforts by Sert and Doxiadis to define the related inter-disciplinary fields of urban design and ekistics, respectively.

EKISTICS at Harvard

The growing scope of DA's operations, in conjunction with the UN's expanding involvement in human settlements, intensified the need for Tyrwhitt's monthly *Bulletin*, and also multiplied the amount of material she had to review, abstract and compile. For the first year, Tyrwhitt had the help of GSD librarian Katherine McNamara. During the second year, Caroline Shillaber, Librarian of the Rotch Library at MIT, took over. In the third year Tyrwhitt hired a secretary; Yvette Mann, a young French woman who could also help with translation, began work at the end of June 1957.

With Yvette Mann on board, Tyrwhitt felt free to go to Berlin in August, ostensibly to attend an IFHTP conference, where one panel would discuss "under-developed" regions. "I could ... represent Harvard, UN and Doxiadis simultaneously!!," she exclaimed, proud of the many hats she now wore. [SG.103] The real reason for her trip to Berlin, however, was to undergo a "cure" for her debilitating asthma; Giedion had arranged for treatment at a clinic that employed naturopathic methods. Following her three-week treatment, Tyrwhitt and Giedion traveled together to attend a final meeting of the CIAM Council, at which the old CIAM was dissolved This was effectively the end of Tyrwhitt's involvement with CIAM, except for a future attempt to write a *Charte de l'Habitat*, and, later, a history of the organization, neither of which came to fruition.

Back in Cambridge in September, Tyrwhitt began producing an expanded version of the *Bulletin*, which, as of the October 1957 issue, came out in a new format, featured Buckminster Fuller's dymaxion map of the world on the cover, an editorial, and a new title: *EKISTICS Housing and Planning Abstracts*. In her debut editorial, Tyrwhitt put her own spin on the definition of ekistics; she considered it human ecology, "the inter-relation of man and environment, including the systems of human settlement." [E.12] Articles would continue to focus on tropical housing and planning, she explained: "because we think the biggest jobs are yet to be done in such areas. But we shall be increasingly alert to developments throughout the world." The circulation of *EKISTICS* was increased to include university departments, research institutes and other groups willing to send material to Tyrwhitt on an exchange basis, an arrangement similar to the one she had with Bauer in the days of APRR's Information Service. This material would later be collected in the library Doxiadis was building in Athens, indexed according to his own classification system—just as Tyrwhitt had done for APRR's library. Tyrwhitt was realizing her vision for APRR through her work with Doxiadis.

EKISTICS began at an opportune time: the October 1957 Sputnik launch spurred efforts in the West to create a new information infrastructure to promote scientific innovation in an array of new, multi-disciplinary fields. Thanks to her new secretary—and Yvette's husband Lawrence, a graduate student in the Harvard planning program—Tyrwhitt found the job of producing the expanded bulletin "greatly lightened." [SG.104] Lawrence had worked with Ruth Glass, in London, and sought out Tyrwhitt because of his interest in Geddes.[2] Yvette and Lawrence Mann compiled *EKISTICS* under Tyrwhitt's direction through July 1959, while Lawrence also worked as Meyerson's research assistant. As a result of Meyerson's arrival at Harvard Tyrwhitt felt "sure things will gradually improve a lot both in the planning department and in the inter-connection of planning and architecture." [SG.105] Meyerson and Mann both joined Tyrwhitt as supporters of Doxiadis and leaders in what became the ekistics movement.

---

2   Personal communication with Lawrence and Yvette Mann, April 2008.

Figure 14.3 **In 1957 the monthly bulletin Tyrwhitt produced for Doxiadis was expanded, renamed EKISTICS, and "re-branded" with a dymaxion world map on the cover**

Urban Design as a New Discipline

Sert relied on Tyrwhitt to organize several meetings in 1958 to prepare for UD 3, one aim of which was to solidify support for the new Urban Design degree program, which hadn't yet been approved. In April 1958, three small panels met for two days and decided that the conference would seek "to arrive at certain principles which can guide the design of large scale residential developments of an urban character." [L.31] A working seminar in November 1958, selected and analyzed six recent projects to discuss at the conference the following April. It was also decided that a fourth UD conference would include a similar comparative analysis—a sequence similar to CIAM methods.

Additionally Sert relied on Tyrwhitt to relate the discussions in the UD Seminar: The Human Scale—which was conducted by Giedion in cooperation with Sert and Sekler in the spring term of 1958—to preparations for the UD conference. In its second year the seminar broadened its focus from comparative studies of different systems of proportions, and their use in antiquity, to include studies of recent large-scale residential projects. Tyrwhitt actively participated in the seminar, in addition to teaching her own courses; she was knowledgeable about the cases, all of which had been analyzed at CIAM congresses. The seminar also took advantage of having students from Japan, China, and India to bring non-Western perspectives into the discussion. Tyrwhitt helped build the seminar as a platform to weave the CIAM line of thought into a broader discourse on urban design that incorporated a new, global dimension.

In October 1958, the beginning of her fourth year at Harvard, Tyrwhitt's heavy workload became at least psychologically easier to bear when Sert quietly promoted her to Associate Professor. ("It was indeed high time and I would have appreciated if Sert would not have been SO discrete," Giedion commented.) [SG.106]

That spring Tyrwhitt supervised the work of students who prepared case studies for discussion at both the third UD seminar and the conference. GSD alumni had selected the cases, which did not appeal to Giedion: "About the town planning conference and its program ... I don't find the comparisons too exciting," he commented. [SG.107] He was, however, interested in having the seminar discuss a competition for a new Toronto City Hall and Civic Square, on which both he and Tyrwhitt had written. In her article—on the organization of the Civic Square portion of the entries—Tyrwhitt expressed concern that competitors and judges "seemed to pay very much less attention to this part of the problem" (1959: 55). She pointed out the "part the civic square should play as a coordinator of the civic hall complex with the surrounding city scene." Just as she had been instrumental in foregrounding the CORE in CIAM debates, she focused the attention of her students, most of them architects, on the importance of the form of open space— the "space in between" where interactions between "you" and "me" take place— as a key component of urban design.

Sert's efforts to build support for urban design education paid off in May 1959, when the Harvard Corporation approved three new Master degrees in Urban Design, the first in the US, to start in Fall of 1960; in Sert's words, "physical planning in our cities," finally emerged (in Design 1959). He counted on Tyrwhitt, who remained on the faculty until they both retired in 1969, to accomplish that objective. "The ten conferences that ensued, although they bore the stamp of the vision of ... Sert, were organizationally the work of Jacky," confirmed William Doebele, who joined the faculty in 1958 and became Associate Dean in 1965 (1983: 440). "After five years of UD conferences, enough common ground had been laid out for Dean Sert to open Harvard's doors to a formal program, an enterprise in which Jacky naturally had an indispensable organizing role."

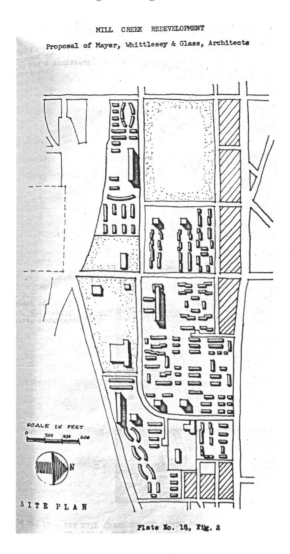

MILL CREEK REDEVELOPMENT

Proposal of Mayer, Whittlesey & Glass, Architects

SITE PLAN

Plate No. 18, Fig. 2

Figure 14.4   Tyrwhitt supervised the work of students who prepared case studies for discussion at the urban design seminar and third Urban Design Conference in 1959

Ekistics as a New Discipline

In tandem with her work with Sert Tyrwhitt also supported Doxiadis's efforts to build an institutional base for ekistics as a new discipline; the establishment in 1958 of the Athens Technological Institute (ATI), the first school to teach ekistics, was particularly significant. "Technology—the modern international frontier" framed her comment in the August 1959, issue on the connection between the efforts to institutionalize education in ekistics and urban design at ATI and Harvard, respectively. [E.13] "The advancement of the technology concerned with the science of EKISTICS has come about in three ways: by education in new techniques; by their refinement and development and by the perfection and development of new machines." She presented the new two-year graduate program in Ekistics that ATI began offering in January 1959, and the Urban Design Conferences sponsored by GSD since 1956, as examples of education in new techniques. Both programs provided platforms for advancing "theoretical and practical knowledge" and considering "the interplay of theoretical ideals and practical schemes."

Doxiadis began to exert more control over *EKISTICS* beginning in January 1959, when the printing of the bulletin moved to Athens; issues now often included a lead article by him and case studies of his firm's projects. Tyrwhitt was insistent, though, to maintain its integrity the journal needed to not be seen as a marketing tool for DA. Even as the next six issues featured lead articles by Doxiadis laying out the basic framework of the "problems and science of human settlements" (the subtitle of the bulletin as of May 1959), Tyrwhitt's editorials highlighted multiple points of view and emphasized that there wasn't an orthodox approach. She wrote in April:

> Certainly the "science of human settlement" has gone no further than the tentative putting forth of models of sequential occupancy and the natural science of description and classification. It is only through feelers among available material on "settlement-development" that the scope and content of Ekistics as a discipline can be realized: and this is one of the key reasons for the existence of this bulletin. … [T]he articles … may provide some hint for a field of study which is beginning to emerge. [E.14]

Face of the Metropolis

From her perch at Harvard and an editorial position at *EKISTICS,* Tyrwhitt had a unique perspective on the interactive nature of programs launched by the UN, GSD, and Ford Foundation. Her June 1959 editorial observed: "Almost as though it were following the lead of the recent United Nations Seminar on Regional Planning in Tokyo, ACTION (The American Council to Improve Our Neighborhoods) recently decided to expand its activities from neighborhood improvement to the problems of metropolitan areas." [E.15] Meyerson, who headed ACTION from 1952–57, had attended the UN seminar, where "The Shape of Our Cities" exhibit was on display. Another participant in the seminar was Paul Ylvisaker, of the Ford Foundation, a funder of ACTION. Meyerson hired Tyrwhitt to work with him, beginning in the summer of 1959, on a new ACTION project: an urban design study that was published as *Face of the Metropolis* (1963); the influence of "The Shape of Our Cities" on that publication is evident. Tyrwhitt and Sert were then considering revising "The Shape of Our Cities" for publication under the auspices of CIAM. [T.82, 83] But Sert lost interest in that project, and it's likely that Tyrwhitt recycled some of the material in *Face of the Metropolis.* Tyrwhitt did not, however, receive full credit for her work. Meyerson acknowledged in the text, but not on the cover, that Tyrwhitt shared with him "the responsibility for organizing the book" (1963: 9) In correspondence with the publisher, he referred to her as his co-author. [T.84]

Mediating Giedionese

Throughout her time at Harvard, Tyrwhitt also worked with Giedion on his two-volume book *The Eternal Present: A Contribution on Constancy and Change.* Tyrwhitt completed most of her translations of Giedion's German text for *Volume I. The Beginnings of Art* (1962) by July 1957, but then she was too busy to do any more, and he had to rely on other translators, none of whom were satisfactory. Tyrwhitt made time to work with Giedion to revise those translations during the 1958 spring term when he was at Harvard but Giedion continued to struggle with the text. In October, he pleaded with her: "You have no time to work on ANY THING! I know. But you could help very much if you could go through the whole manuscript & ... make it more fluent & send me from time to time your proposals!!!!!!!!!!" He added an emotional appeal: "I am working on things we saw together [in Iraq] .... but I would like to write it with you: for looking at things it was the most intensive time, we ever had together." [SG.107] Tyrwhitt relented and met his February 1959 deadline. Giedion assured his nervous publisher that Tyrwhitt would "correct the different versions with me so that there is only one voice through the whole manuscript." [T.85] In August Giedion told Tyrwhitt that there were still "certain holes" in his text. "I will send you things in bits. Just as we did it last term at Harvard." [SG.108] Tyrwhitt finally drew the line, and stopped working with Giedion after December 10. "I was 'at the end of my tether,'" she later acknowledged. "I was both too tired and pressed for time to go into long explanations." [SG.109] She had to pack and sort out what would have to be done while she was away from Harvard for the next six months. After spending Christmas with her family in Wales, Tyrwhitt traveled to Bandung, Indonesia, to do something entirely new.

Chapter 15

Synthesis:
Urban Design and Ekistics

A Planning School in Asia

The reason Tyrwhitt traveled to Bandung Indonesia in December 1959 was to become the second Harvard faculty adviser in residence to a new School of Regional and City Planning, which opened in September 1959 as a division of Bandung Institute of Technology. The school was founded with the joint support of the UN Technical Assistance Board, Government of Indonesia, and Harvard University Department of City and Regional Planning. But the idea for the school is crystallized in the context of the transnational networks that Tyrwhitt had helped establish through her work with APRR and SPRRD, and as consultant to the UN. Tyrwhitt's connection to the project dates to the UN Seminar on Housing and Community Improvement in Asia and the Far East in New Delhi in early 1954, where UN ECAFE formally endorsed a proposal by Professor Ir K. Hadinoto, the chief Indonesian delegate to the Seminar, to establish a planning school in Southeast Asia. The implementation of this idea in Bandung played out over the next decade against the backdrop of the cold war and decolonization.

Background

Indonesian nationalists established a central planning bureau within the Ministry of Public Works and Reconstruction in 1946, but implementation had to wait until Dutch recognition of Indonesian independence in December 1949 (Roosmalen 2005, Webster 2011). There was a shortage of knowledgeable professionals to take over from the departing Dutch because Indonesians' access to higher education had been restricted during the colonial period. However JK Thijsse (1896–1981), director of the central planning bureau, was an advocate for education in planning in Indonesia. After the transfer of sovereignty, Thiijsse joined the faculty of a new School of Architecture, which occupied the buildings of the former Bandung Institute of Technology, as a division of the new University of Indonesia (UI). Shortly afterwards, Thijsse was appointed to the Crane Mission of Experts. In 1951, Hadinoto became chief planner in the Ministry of Public Works. He soon explored setting up a planning course in the School of Architecture at Bandung with help from British town planner Clifford Holliday, a UNTAA adviser in Djakarta, affiliated with the University of Manchester. That partnership didn't materialize. Subsequently, Hadinoto became director of the UN-sponsored Regional Housing Research Center, which opened in Bandung in 1955.

Meanwhile, Tyrwhitt spent two days with Thijsse—who was affiliated with CIAM and IFHTP—in Bandung in 1953, in preparation for the UN seminar in New Delhi, in which he served as a discussion leader. [SG.109] She also met with Holliday in Djakarta. He asked Tyrwhitt for a reference for one of her former students applying for a job to become his replacement in Djakarta. Instead, she recommended another former student, Kenneth Watts, who was then working in Singapore. Watts got the job and began his assignment in 1956. [T.86, 87]

That year, President Sukarno signed an agreement with the US Government and foundations to bring in teams from American universities to train Indonesian faculty members. The International Cooperation Administration (ICA) contracted with the University of Kentucky (UKy) to develop an engineering and scientific research and training center in Bandung, which became known as the Institut Teknologi Bandung (ITB), and included the UI departments in Bandung. UKy planned to send Indonesians to American universities for advanced training; they would return to take positions as ITB faculty. But the political situation was extremely volatile; there were periodic protests calling for sending the UKy team home (Rice 1969).

University of Indonesia leaders were initially skeptical about the idea of a school of planning, perhaps remembering problems perceived with the earlier proposal. But they were receptive to the idea of a partnership between Harvard and the UN to start up the school as a pilot project for the global system of research and training institutes Weissmann envisioned. Weissmann's idea grew out of the recommendations of the 1954 New Delhi seminar, which Tyrwhitt reported on in *EKISTICS* in February 1957. [E.16] Meyerson, who served as senior advisor to the Harvard/UN partnership, asked Tyrwhitt to be involved when the project was first discussed in spring of 1957, even before he joined the faculty. Ir Soefaat, one of Meyerson's students at the University of Pennsylvania, upon graduation that year was returning to Indonesia to become chief planner at the Ministry of Public Works. Soefat soon became a collaborator with Watts and Hadinoto on the planning school initiative (Watts 1997).

William Doebele became a key member of the team when he joined the Harvard faculty in February 1958. That spring, Watts traveled from Djakarta to Cambridge to discuss the Harvard/UN partnership. Following his visit, Tyrwhitt galvanized her network in Southeast Asia to do reconnaissance; she wrote to one of her students, Gwen Druyer, then in Australia with her future husband Gordon Bell, to let her know confidentially that Harvard was "about to undertake the sponsoring (in quite a big way) of a new SE Asia School of Planning (confidential as of yet)," and asked her "to make certain contacts in this connection" during her travels. [T.88]

Throughout 1958 Meyerson, Tyrwhitt, and Doebele developed the Bandung project in consultation with Weissmann and Hadinoto and Soefaat, who served as Indonesian advisors to the UN-Harvard team. The key, Doebele recalled, was "to attack head-on the question of education for a kind of planning that does not yet actually exist" (1962: 97). Doebele and his wife accompanied Meyerson to Bandung in January 1959; Meyerson stayed for a month, and the Doebeles stayed on for the first year. Sert and Tyrwhitt were to go the following year, but she ended up going on her own, for six months.

Doebele had warned in June 1959: "everything which could go wrong has done so." [T.89] Compounding the political and logistic difficulties in Indonesia, in 1959 the UN secretariat reorganized: the Technical Assistance Administration (TAA) was folded into a new Bureau of Technical Assistance Operations (TAO), within a merged Department of Economic and Social Affairs. In addition, the UI Faculty of Architecture regained its independence, and was renamed Institute of Technology Bandung (ITB). When Tyrwhitt arrived, however, she took advantage of her previous experience in the region, and the groundwork done by the Doebeles and Watts. She told Giedion: "It is so easy here! First I am doing something I find interesting and something I have done many times before: training young and intelligent people in the elements of the practice of urban planning. The simple techniques of conducting a survey and interpreting material and determining the main lines of a plan. ... It's also extremely interesting for me to see how to vary the same basic techniques to meet the conditions of this country—and the 40 students are both extremely eager and ... very quick in the uptake." [SG.112]

Tyrwhitt found life in Bandung "peaceful and without cares." She was "able to work at a slow tempo in an ideal climate being waited on hand and foot," and her life became even more pleasant when her erstwhile travel companion Delia arrived. [SG.110–112] In March she assured Sert, "As far as I know this is the most efficient, and realistic and economic assistance that has yet been given in setting up a department which should be able to run (at a high level) under its own power within 5–7 years from the start." [T.90]

Before leaving in July 1960, Tyrwhitt reported on what she had accomplished and outlined the next phase of work. The main tasks during the startup period (1959–1961) "have been and still are: to get good students enrolled and to teach them; to get adequate building space and a good library; to frame the long term curriculum and to enroll a first class Indonesian faculty to be trained abroad." [T.91] While the four-and-a-half-year curriculum was already settled in principle, the resident faculty had to make some modifications for the first few years until the trained Indonesian faculty returned to Bandung. Tyrwhitt taught an introductory lecture course; and she co-taught with WJ Waworoentoe two physical planning studio courses. Students undertook field surveys of the use of house plots in two nearby villages, which then formed a basis for the determination of plot sizes in the plans they would make for these areas; they assumed the construction of all new houses would be undertaken by the people themselves. [E.17]

To guide these courses, for which there were no available written materials, Tyrwhitt devised a cost-effective strategy to publish two provisional texts, one of which was written by Watts on survey techniques (and summarized in *EKISTICS* in May 1962). [E.18] Tyrwhitt and Waworoentoe organized supervised paid summer internships for the students to supplement class work with field experiences. Thanks to their supervision, the placements were considered successful and were repeated the following year. Tyrwhitt also took particular interest in setting up the school's library; it was her "special pet." [T.92]

Looking ahead, Tyrwhitt thought: "The end of 1963 probably represents the first moment that the school could safely make its existence widely known in South East Asia." To celebrate this final phase, Tyrwhitt proposed that a long discussed UN Seminar on Planning, Training and Research in South East Asia be held there in January 1964. "It is reasonable to believe that by then the need for UNTAO/Harvard guidance of the school's development could be considered to have reached its end. The whole project should be ripe for a new stage of development: the actual setting up of a SE Asian Center for advanced training and research in regional planning." Meyerson wrote Tyrwhitt before she left Indonesia in July: "The UN is very pleased with all that you have done and are only astonished, as we all are, at your energy." [T.93]

Tyrwhitt reported on the new school in the May 1960 *EKISTICS*, which was the first issue to feature original articles, including one by Jacob Crane. She noted that Crane "considers the two most hopeful areas of achievement [in the field of international technical assistance] to lie in the promotion of technical education within the country receiving technical aid and the development of local building materials ... ." [E.19] Her work establishing the school and her article about it exemplified just those achievements.

Institutionalizing the ITB School of Regional and City Planning

After departing from Bandung in July 1960, Tyrwhitt didn't return immediately to Cambridge. She stopped in Athens to touch base with Doxiadis, then spent two weeks with Giedion in Switzerland, and put in a day at UN headquarters. Over the next decade Tyrwhitt retraced variations of this route, becoming a global citizen, dividing her time between GSD, Doxiadis' Athens Center for Ekistics (ACE), and the UN ECAFE region. In this way she facilitated the cross fertilization of planning ideas in every institutional setting she engaged with, and helped build an international planning academic community.

Back in the US, Tyrwhitt collaborated fruitfully with UN personnel in New York, and with Meyerson and Doebele at Harvard in order to help the ITB program succeed: guiding the work of succeeding resident advisors and strengthening ties connecting the nascent school to Harvard and the UN. In her final report, Tyrwhitt had expressed her concern about "the whole question of the organization of a worthwhile and financially sound programme of research, both to make full use of the newly developed skills of returning Indonesian faculty members and to provide them with additional income." [T.91] At a meeting at UN headquarters in October 1960 she reiterated the plea she had made previously to restore funding for a second UN expert to deal specifically with the research program—a post that was "inadvertently omitted in the Government's request for TA in 1961–1962." She hoped the UN—which could only respond to a host country's request for aid—could diplomatically suggest that the Government include this position in its 1962 program budget. [T.94]

In October she also recommended young Indonesians suitable for UN fellowships available to train future faculty, and agreed to develop their study plans in consultation with Doebele. Tyrwhitt further proposed that all of the Indonesian

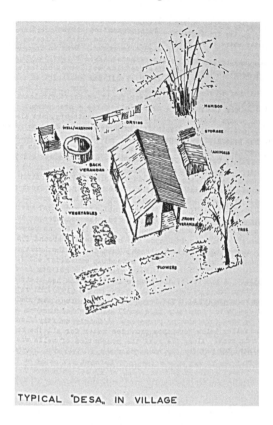

TYPICAL "DESA„ IN VILLAGE

Figure 15.1 **Tyrwhitt taught the first planning studio in Indonesia in which
students applied survey techniques in the field. Students used
their surveys of house plots in nearby villages to establish the
plot sizes in the plans they made for those areas; assuming
that all new houses would be built by the people themselves.
These simple, down to earth techniques "are universal and
fundamental to regional development everywhere in the world,"
she wrote in an article about this project in Ekistics [E.19]**

faculty members in North America on UN fellowships meet at Harvard in the
coming spring to review their work with Harvard staff connected to the project.
The UN agreed to sponsor the meeting if Tyrwhitt convinced other funding
organizations to send their fellows too.

Due to the rapport she established with Indonesian students and faculty she had met
while in Bandung, and her network of former SPRRD students working in Southeast
Asia, people often reached out to Tyrwhitt for informal advice and assistance. In March
1961, the chairman of the newly formed Indonesian Planning Students Association
wrote asking for help in making contacts with planning students and professional

associations in the US and elsewhere, and for help in setting up their own professional association. Tyrwhitt replied that the best persons to contact were the future Indonesian faculty now abroad, as this problem "must be worked out in Indonesia," she assured them that their letter would be "thoroughly discussed" by all future faculty, UN officials, and Harvard faculty when they met at Harvard in May. [T.95]

Tyrwhitt opened that conference by hosting a party in her apartment. In the course of the two day meeting a variety of issues were raised, including the need to quickly produce research publications, the problem of limited data availability, and how the future faculty members could most effectively work together. "It was generally agreed," the minutes noted, "that the establishment of the profession was one of the most important and difficult issues that will arise once the first graduates have been produced. This was a matter which both the faculty and the students should cooperate in solving ... in a way that would be most useful in the culture and institutions of Indonesia." [T.96] Tyrwhitt foregrounded that issue for the participants, in line with the importance she attached to institutional development as a key aspect of the transformation under way in the country.

In July 1963 she was asked by the ITB School of Architecture—which housed the planning program—to act as its representative in the USA. The invitation read:

> We would like to be under the patronage of someone not only with authority in architectural and university circles, but with knowledge about the situation of and a warm heart for our school in Bandung. Such a person, dear Professor Tyrwhitt, we thought you to be exactly and without reserve. [T.97]

Tyrwhitt agreed to represent the group, although she contemplated retiring in 1964.

When Meyerson left Harvard in September 1963, he entrusted the significant investment he had made in the Bandung project—and his reputation—to Tyrwhitt, who, along with Doebele, replaced him as UN Senior Advisors. Due to political turmoil, though, Americans were forced to leave Indonesia, and in 1965 Indonesia withdrew from the UN temporarily. Tyrwhitt stayed in touch with the project through connections with the new Australian advisers, and through her UN contacts. When Tyrwhitt returned to Southeast Asia in 1967 to advise on the establishment of a new planning program in Singapore, she was pleased to find the planning program in Bandung thriving. The program's continuing success testified to the dedication of the Indonesian faculty and the resilience of the institution that Tyrwhitt and the rest of the UN/Harvard team helped them build.

Teaching Urban Design at Harvard 1960–64

While Tyrwhitt was in Indonesia, her contribution to Sert's vision for GSD was sorely missed. Isaacs had proposed dropping the environmental design studio required of all first year students, which had already been reduced from a full year to a semester. This move reflected a broad trend in urban planning education in the US towards greater concern with social and economic factors than with physical, spatial, or aesthetic considerations; and Isaacs's insistence on "a high degree of autonomy with collaboration at the will of the instructors" for the department he directed. Tyrwhitt told Sert she opposed this: "It is the one chance *all* the planning students have to become acquainted with some general principles of design and the one chance all the architecture students have to take the wider needs of a community into consideration. I think both groups of students would lose much by splitting the course in two." [T.98, 99] Sert agreed, but reached a temporary compromise, excusing more advanced planning students entering in 1960 from the environmental design studio, and charging the faculty to come up with a better long-term solution. In this context, Sert wanted Tyrwhitt to teach environmental design in the fall of 1960 in order to provide continuity. [T.100] Tyrwhitt "officially" joined the urban design faculty the following year.

By June 1962, when Tyrwhitt reported on urban design education at Harvard at a conference organized by the American Institute of Architects (AIA) and the Association of Collegiate Schools of Architecture (ACSA), most of the students

Figure 15.2   The cover of the report of the Fifth Urban Design Conference, held in April 1961, featured this photograph showing Tyrwhitt in her characteristic position, at the center of attention but behind the prominent man—Jose Luis Sert—in charge

enrolled in the Harvard urban design program were architects. Moreover, it had "become less and less practicable to organize" opportunities for collaborative work on an urban design problem beyond the first term environmental design studio (Tyrwhitt 1962: 101). The rift between planning and design faculty and students at GSD deepened over time.

Tyrwhitt explained at the AIA/ACSA conference that Harvard's urban design program operated on two scales: the macrocosm and the microcosm. "The first is a frame of reference, conceptually sensed but not necessarily visually apparent at any one moment. The second is directly concerned with what is physically visible at the human scale ... with the design of variant elements within the conceptual system" (1962: 100, 1966a: 122). Tyrwhitt felt more confident theorizing about the former than the latter. She became immersed in the problem of how large-scale urban forms could provide a pattern "for living in equipoise within a rapidly changing environment," when the urban design studio and the Sixth Urban Design Conference in 1962 addressed the theme: Designing for Inter-City Growth. By then, several models for metropolitan growth had been identified as alternatives to the sprawl spurred by highway construction and post-war suburbanization. Tyrwhitt coordinated the urban design studio with conference panels that studied four of these alternative patterns: new towns, characterized as *dots* in the areas or growth between cities; inter-city corridors, visualized as *lines* radiating from the city center; concentrated peripheral growth, considered as *rings* growing around the city; and rationalized sprawl, dubbed *rugs*. [T.101] Those mid-20th century alternatives continue to frame early 21st century debates about planning for "smart growth," "new urbanism" or "sustainable development" versus unconstrained market driven processes.

To conceive of the macro-design of large metropolitan regions posed a new challenge, entirely different from the micro-design of the core of the city or elements of urban renewal projects, which previous UD conferences had addressed, and which could be approached with design principles derived from architecture (Gutheim 1962). An approach that Tyrwhitt found particularly compelling was the idea of collective, or group form conceptualized by Fumihiko Maki, a GSD alum then teaching at Washington University, who taught the urban design studio at Harvard in spring term 1962 and subsequently joined the faculty. Tyrwhitt featured his thoughts on "linkage in collective form" in her report on UD 6 in *EKISTICS*. [E.20]

Maki was part of a group of young Japanese architects and designers known as the Metabolists, who understood the connection between the roots of modernism and tenets of Japanese aesthetics, and were using them to invent a modern Japanese urbanism and architecture. The Metabolists's solutions to the problems of rapid urban growth, while visionary, were actually grounded in the reality of the chaotic expansion of Tokyo—then the densest and fastest growing city region in the world—where regional planning based on British principles—green belts—had proven ineffective (Ross 1978: 23). Tyrwhitt's association with Maki, with whom she co-taught the urban design studio at Harvard in 1962–1964, and with whom she remained close afterwards, deepened her appreciation of the Metabolists's proposals, which in trying "to encourage the active metabolic development of our society" had an affinity with Geddes's ideas (Kurokawa 1977: 27).

Another reason Tyrwhitt was attracted to the Metabolists is that their ideas resonated with general systems theory, as promulgated by two thinkers she admired: Kenneth Boulding and Buckminster Fuller. Metabolist proposals were also analogous to Doxiadis's studies of the City of the Future (COF)—conducted by a team of researchers at ACE— which she collaborated on in 1961–1962. Tyrwhitt drew on all of these concepts as she refined her hexagonal diagram as a theoretical model for a region that could absorb rapid growth without destroying existing communities.

Tyrwhitt presented this model in a paper titled, "Shapes of Cities That Can Grow," at AA in June 1963. Her premise: "when cities, extensions of cities or new housing areas have to be built, we all turn to a set of models ... and consciously or unconsciously judge from the standpoint of some utopian ideal," usually from the

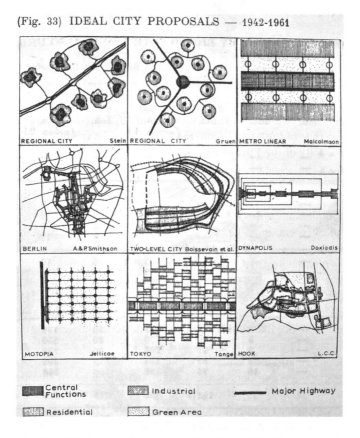

(Fig. 33) IDEAL CITY PROPOSALS — 1942-1961

Figure 15.3 **Tyrwhitt's contribution to the City of the Future project included a review of 20th c. utopian plans that incorporated possibilities of growth and change**

Figure 15.4   Tyrwhitt addressing a meeting at ACE including Hasan Fathy (left at table) and Constantinos Doxiadis (right), discussing the City of the Future Project

Source: © Constantinos and Emma Doxiadis Foundation

past, for example, the garden city. The global spread of urbanization signaled the emergence of a new era, and called for a new urban ideal. Echoing Boulding—whose "The Death of the City: A Frightened Look at Post-Civilization," she had excerpted in *EKISTICS*—Tyrwhitt declared: "The 'city' in its former sense no longer exists. What we now live in has been called a conurbation (Patrick Geddes), ... megalopolis (Jean Gottman), dynapolis and ecumenopolis (Constantinos Doxiadis)." [E.21] She praised Kenzo Tange's 1960 mega-structural plan for Tokyo, which relied on transportation infrastructure to provide the central organizing framework. She reasoned that strategic public investment in transportation systems could lead to a desirable urban form, assuming public control over land use. Citing a proposal that attached development areas to transportation corridors across Britain and into Western Europe, Tyrwhitt presciently concluded: "This system ... is the kind of scale on which we must plan for the next two generations of population expansion" (1963: 87–88, 96)

The "tentative composite diagram" Tyrwhitt presented showed schematic residential zones, administrative centers, schools and recreation areas, and industry linked by a limited access highway system with recurrent three point junctions at 120 degrees. Key services were clustered in specialized centers, and school and recreation areas would be traffic-free. To demonstrate "how such a system might be put in practice as a basis for planning," Tyrwhitt chose the East Anglia region of Britain, because of its relatively flat topography (1963: 99).

Tyrwhitt *wanted* her model to provoke debate, and she received some harsh criticism from those in the audience who took her diagram "a little too literally." (1963: 101). Perhaps that's why she didn't develop her model further. Nevertheless, she encouraged the next generation of urban planners and designers to envision new metropolitan forms. The idea of regional planning to guide urban growth into a polycentric constellation of compact centers has since been widely accepted as a sustainable pattern of development.

Doxiadis and the Delos Symposion

Tyrwhitt had prepared "Shapes of Cities That Can Grow" to serve as a resource for the first Delos Symposion, in 1963. She went from London to Athens to help Doxiadis and his colleagues at ATI get ready for this gathering, which took place July 6–16 on board a ship cruising the Aegean Sea. Tyrwhitt and Doxiadis first discussed such a meeting in 1958. She suggested that Doxiadis invite "some of the architects who are really interested in design for 'developing countries' to discuss some of their common design problems—coming with work in their hands." Doxiadis had a similar idea: "He thought of inviting a number of people to discuss 'ekistics'— two days in Athens and 5 days on Delos," she told Giedion. "It would not be CIAM, nor pretend to be, but it would be one of the children CIAM has spawned." [SG.113] Tyrwhitt helped Doxiadis focus his ideas about what became the Delos Symposion as she gradually became a key member of his circle. She likely advised Doxiadis on his proposal to Ford Foundation in 1959 requesting support for an "International Symposion on Urbanism"—which was rejected. [F.4] Doxiadis used the Greek spelling "symposion," rather than the Latinized "symposium," to convey an informal meeting with free exchange of bold ideas. [E.22]

Doxiadis continually urged Tyrwhitt to take on a larger role in the division of ATI that included the Graduate School of Ekistics (GSE) and research projects, which he wanted to expand as the Athens Center for Ekistics (ACE). In 1960, Doxiadis requested funding for ACE as part of a proposed "master plan" for multi-year support submitted to Ford Foundation. [F.4] Doxiadis knew that having Professor Tyrwhitt of Harvard as a key member of the ACE team would strengthen his proposal; he prevailed on her to work in Athens (on the COF project) during January 1961, and to join him in New York for a meeting with the Foundation in early February. [SG.114] That prospect certainly figured into her calculations when, while working for Doxiadis in Athens in summer 1961, Tyrwhitt decided she would retire in Greece. The following summer she began looking for a site to build a house, and by the end of the 1962 she succeeded in purchasing her first hectare of land on a hillside called Sparoza outside of Athens. She also took on a short UN consulting job in Gambia that winter to help pay for her new house.

Tyrwhitt assumed a wider range of responsibilities for Doxiadis in Athens beginning in the summer of 1962. To begin with she edited and organized the production of his book for the National Association of Housing and Redevelopment Officials (NAHRO)—an important project because Ford funded it. Doxiadis and his associates leaned heavily on her editorial skills for many of their publications in English. [F.4]

She also joined the planning committee, which was charged with organizing what by then was being called the Delos Symposion, which Doxiadis was able to sponsor thanks to funding from a Greek donor. [E.23] Doxiadis surely relied on Tyrwhitt's organizational skills as well as her London, CIAM, Toronto, Harvard, and UN contacts. [E.24] The guest list for the first Delos Symposion—34 prominent academicians, architects, businessmen, and government officials from 15 countries— is studded with people with whom Tyrwhitt was connected.

Doxiadis had a specific vision of how he wanted to conduct the meetings, and Tyrwhitt was shocked, in April 1963, when she received a copy of his program: "No agenda, no minutes, no report—or words to that effect. But this cannot be entirely true if anything more than a brief statement of banal generalities is to be produced," she told Giedion. Not sure of the role she was expected to play, she hoped it would not be "too exacting since the whole voyage looks delightful." She was also unclear about what Doxiadis expected to come of it. "But I like the man and would not like to see the thing a fiasco," she said, explaining her willingness to comply with Doxiadis' directions. [SG.114, 115]

Tyrwhitt discovered in June that Doxiadis intended her to play a very demanding role, one crucial to the realization of his vision of an informal symposion. Acting as the only secretary, she was to capture the spirit of the discussions, rather than provide a detailed account of the proceedings. [E.24] This meant that she was continually taking notes, and worked "through the nights to get the minutes on the breakfast table." [SG.116] Nevertheless, Tyrwhitt declared: "I enjoyed it, and the prima donnas were remarkably relaxed and cooperative: indeed everyone had a good time and decided … that it was a good idea to make a 'Declaration of Delos.'" [SG.117] Producing the declaration took place over the course of three meetings, during which Tyrwhitt faced the formidable intellectual challenge of distilling the differing points of view expressed by specialists from many different fields. But she was not invited to join those who lined up in torchlight in the ancient theater of Delos to take part in the ceremonial signing of the document. She made her crucial contribution to the process as a member of Doxiadis' staff.

Tyrwhitt expressed the ideas she had hoped to contribute to the first Symposion in her editorial introduction to the June 1963 issue of *EKISTICS*. She argued that in antiquity, "the growth and change of human settlements exhibited an obvious but unconscious integration" of economic, social, political, administrative, technical, and aesthetic factors. However with urbanization and industrialization urban planning in the Anglo American world had become "totally divorced from the disciplines related to art, and all attention has been concentrated upon the determination of social and economic goals and (more recently) the processes of decision making." Urban design courses began to appear when the "fallacy of this approach" was recognized, but in many schools, urban design "continues to mean a simple extension of normal architectural studies." In contrast: "Ekistics places the disciplines related to art on an equal footing" with social and technical sciences. She championed the article authored by a team of young Japanese planners and architects associated with the Metabolists, whom she thought, were "steeped in the full range of ekistic studies." Their views were in touch with the realities "of our age of rapid change, multiple dimensions and 'vision in motion.'" [E.25] She commissioned the translation of the Japanese team's article, "Theory and Methods of Urban Design," as her "indirect contribution" to the first Delos Symposion.[1] [SG.118]

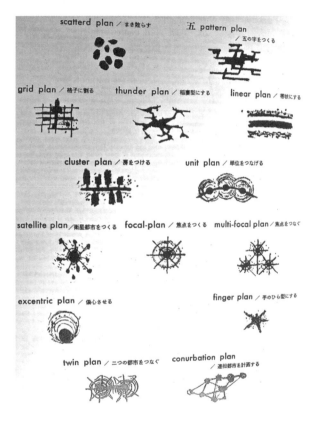

Figure 15.5 Model patterns of urban form analyzed by a team of Japanese architects in an article on City Design in *Kenchiku Bunka*, that Tyrwhitt had translated for Delos I

---

1 Tyrwhitt had Shun-ichi Watanabe translate the article, first published in the November 1961 issue of *Kenchiku Bunka*,which was authored by Teiai Ito, Hidemitsu Kawakami, Shin Isozaki, Michimi Morimura, Koichi Sone, Asahi Tsuchida, Takehisa Odera, Norio Kobayashi, Reiko Tomita and Yasuyoshi Hayashi.

World Society for Ekistics and the UN

"The Delos Symposion was a big experiment," Doxiadis told the top staff associated with this project, including Tyrwhitt: "The moment is ripe to study what has been achieved, how we are going to continue it." The Declaration of Delos called for establishing "a new discipline of human settlements," and Doxiadis believed there was a consensus to call it Ekistics. He proposed forming The Group of Delos, and calling members Delians. This group and the Delos Symposion "should act as the informal catalyst of all the important action to be taken in the field of Ekistics around the world." The group Secretariat would be located within ACE. *EKISTICS* "will be the unofficial scientific organ of all the Delos information." Doxiadis charged Tyrwhitt and Demetrius Iatridis with proposing how to organize the Group of Delos and its Secretariat, and Tyrwhitt was to prepare a special issue *EKISTICS* on the Symposion. [D.3]

This assignment enabled Tyrwhitt to promote a connection to the earlier CIAM meetings—which has since become part of the "mythology" of the Delos Symposia. She later recalled that early on Doxiadis had "envisaged a colloquium on the scale of the early CIAM gatherings, and he had vivid memories of being a student observer at the Athens meetings of the congress of CIAM IV, after the members had disembarked from the [ship] … in which they had sailed from Marseilles" (1978: 129) Tyrwhitt told Giedion that Doxiadis had invited him to deliver the closing address at the Symposion to mark the relationship with CIAM IV. [SG.115] Tyrwhitt made this connection part of the "official record" in her preface to "The Declaration of Delos," in *EKISTICS* (August 1963), and in an appendix to the special issue of *EKISTICS* on the Symposion (October 1963): "The CIAM Charter of Athens, 1933, The Outcome of a Similar Effort." [E.26, 27]

Both the *Charte d'Athenes* and the Declaration of Delos were calls to action, Tyrwhitt later explained (1978: 130–131). But while the participants in CIAM IV were "youthful, idealistic and militant architects and artists," the men and women at the Delos Symposion were "mature leaders." Perhaps more importantly, most of the Delians had been active during the optimistic post war reconstruction period; by 1963 they had reluctantly recognized that the transformation they sought wasn't going to happen. Sensing a pending global crisis, the Delians organized as "citizens of a worldwide city, threatened by its own torrential expansion." [E.26] Whereas the *Charte d'Athenes* provided radical architects and planners with a statement of general principles "to push forward in the different countries," the Declaration of Delos provided a diverse group of influential people—who "understood the interdependence of parts of national and sub-national systems on the whole, and parts of the unevenly developing world on each other" (Mead 1978: 284)—with an ethic to push the *United Nations* forward.

Tyrwhitt expressed her growing frustration with the UN in the May 1963 issue of *EKISTICS*, which reported on the first meeting of the UN Committee on Housing, Building and Planning; Doxiadis attended representing Greece, and she attended as an observer. "We hope … that subsequent meetings will give more specific drive and direction, more fire and imagination to future policies, so that the United Nations goals of improving the lot of mankind may not continue to

Figure 15.6   Tyrwhitt and Giedion at the Delos Symposion, July 1965

Source: © Constantinos and Emma Doxiadis Foundation

founder in the morass of alleviating progressive deterioration." [E.28] One of the recommendations resulting from that meeting was that priority in the allocation of UN funds should be given to regional centers for research and training and for exchange and dissemination of experience and information on housing, building and planning—the functions Doxiadis envisioned for ACE.

Transition: From GSD to ACE

Doxiadis counted on Tyrwhitt to realize his vision for ACE, and in July again offered her a "prominent position" there. Having "always been interested in international action at a professional (not a political) level," she informed Sert that she found this new opportunity very attractive. "I would not be leaving [Harvard] if I believed I was letting the Urban Design program down, but it is now starting to run under its own steam." [E.29] Sert responded as Tyrwhitt hoped he would, offering a new three-year appointment as Associate Professor of Urban Design.

Tyrwhitt then requested and received a year's leave of absence. Her plan was to phase into her new life in Greece, combining part time work for Harvard, for Doxiadis, and for the UN. Notably she prepared a background paper on New and Satellite Towns as a UN consultant for a Symposium on the Planning and Development of New Towns to be held in Moscow in August 1964; the paper was also useful for the UD seminar she taught at Harvard in the fall 1963, which studied new towns. The 1969 UN publication based on this symposium captured Tyrwhitt's multiple commitments, identifying her as Professor of City Planning and Urban Design, Harvard University, United Nations Consultant, and affiliated with Athens Centre of Ekistics (Planning 1969: 255).

Before going on leave from Harvard, Tyrwhitt spent her last weeks occupied with the Urban Design conference—the 8th in the series that she had helped Sert launch in 1956 and the last that she organized and documented. The theme of UD 8, "The Role of Government in Shaping the Urban Core," heralded the dramatic shift in the political climate in the US since UD 1. Many of the attendees now had "a voice in the circles of power," Doebele remarked: "We are now being asked to develop quickly in this country a set of fundamental urban design principles which are meaningful, sound and reasonably universal." But there was only a small group of people—he named Tyrwhitt among them—who were "attempting to develop more general theories." Doebele declared that this "attitude of timidity … is simply no longer relevant to the new age. The present situation demands boldness and a willingness to commit to … the chancy business of being wrong." [L.32] Doebele's words captured the spirit in which Tyrwhitt embarked on her new commitment to Doxiadis' experiment. She also looked forward to being free of asthma and living in her house in the warm, dry Greek climate! [SG.119]

Delos 2

When Tyrwhitt arrived in Athens in June 1964 to help prepare for Delos 2, she knew in advance that her role was to serve as sole rapporteur. The Delians had decided that the next few symposia would concentrate on developing the contents of the science of human settlements—ekistics. Tyrwhitt was "much happier" at Delos 2, partly because she was in better health and had less work, but mainly because "the whole atmosphere was easier," she told Giedion, who had declined an invitation to attend. "The people had come to know each other. I won't say they had begun to work as a team. … but there was a 'group' rather than a collection of people from here and there." [SG.120] Meyerson attended his first of several meetings, and proposed the formation of a World Association of individuals concerned with human settlements to act as a clearing house for information and generally to institutionalize the group spirit, and to open it up to more people. ACE agreed to host the secretariat, and Panayiotis Psomopoulos, a Greek architect on the GSE faculty, agreed to serve as the group's General Secretary. Tyrwhitt was among the first to join after it was established as a non-profit organization in 1965.

Figure 15.7   Tyrwhitt and Martin Meyerson at Delos 2, July 1964

Source: © Constantinos and Emma Doxiadis Foundation

Tyrwhitt quickly edited the special issue of *EKISTICS* on Delos 2 in order to present it to Ford Foundation president Henry Heald, who conducted a personal review of ATI in October 1964, prior to taking action on Doxiadis' master plan proposal. Heald had arrived in Athens a skeptic, but was so impressed by what he saw that he returned to New York ready to approve a $1 million grant, to be paid over three years and applied solely to ACE. [F.4] This meant that Doxiadis would be able to continue hosting the annual Delos Symposia and expand international programs related to the meetings.

Ekistics Documentation Center: Implementing the Ekistic Grid

After returning to Athens from Moscow in September 1964, Tyrwhitt was surprised to learn that she would not be helping to organize either the "World Association of Ekisticians" or the next Delos Symposion. Doxiadis asked her, instead, to design an Ekistics "Documentation Center." He suggested that "thinking what this should be" would complement her editing duties, since documentation was related to the contents of *EKISTICS*. DA had set up a new computer center that year, and Doxiadis was eager to automate an information service to support the work of both DA and ACE. Tyrwhitt was interested, but felt unqualified. Doxiadis didn't want an expert at that point; he wanted a generalist to think about the big picture of it, and then determine what expertise would be needed. [T.102]

One challenge Tyrwhitt faced was that the Documentation Center had to serve both the DA professionals, concerned with the efficient management of complex projects, and the ACE researchers, engaged in comprehensive analyses of past, present, and future cities. She also had to work with Doxiadis' classification system—represented by his Ekistic Grid—a matrix that allowed cross referencing of a logarithmic scale of units of settlements and basic elements of settlements. Tyrwhitt's job was to make the Ekistic Grid operational.

She began by tasking a group of GSE students with testing the Ekistic Grid as a system for classifying the contents of the journal. She presented the results in the January 1965 issue of *EKISTICS*, in which the Ekistic Grid was first published. The test revealed that the grid was "an excellent vehicle for presenting material, but not sufficiently precise, in its present form, to permit accurate and identical recording of information by different people." [E.30] As Tyrwhitt worked on improving the grid's utility as a classification system, she naturally considered it in the context of the long running discourse on standardization of planning concepts and the use of matrices to describe their inter-relationships with which she had been involved since the late 1930s.

Tyrwhitt presented a revised version of the Ekistic Grid in a talk at AA in June 1965, compared to two previous attempts to design such "grids of inter-relations, each one building to some extent on the last:" Geddes's Notation of Life diagram, and the CIAM Grid designed by Le Corbusier (1965: 10, 12). The CIAM Grid offered "a practical diagram of office procedure," which was also useful for presentation purposes. Corbusier's contribution was to design the system with mechanization in mind. " Geddes's contribution was really his comprehensiveness of taking in the whole process and including the inner and the outer world. Doxiadis' contribution is getting scale into the story … from man to universe." Tyrwhitt used Geddes's diagram "repeatedly, and it does help me to see where I should go." The CIAM grid, however, was basically a checklist, and "there were a lot of things that won't really fit." The Ekistic Grid was also a kind of check-list, she explained. "It is still too complicated and I am now in the process of trying to refine and possibly simplify it and marry it into a computer" so that different users could access information in a useful form. Tyrwhitt continued to refine her comparative analysis of these three visual schemata over the next decade, demonstrating that ekistics was grounded in Geddes's synoptic evolutionary perspective and analogous to the CIAM approach to urbanism.

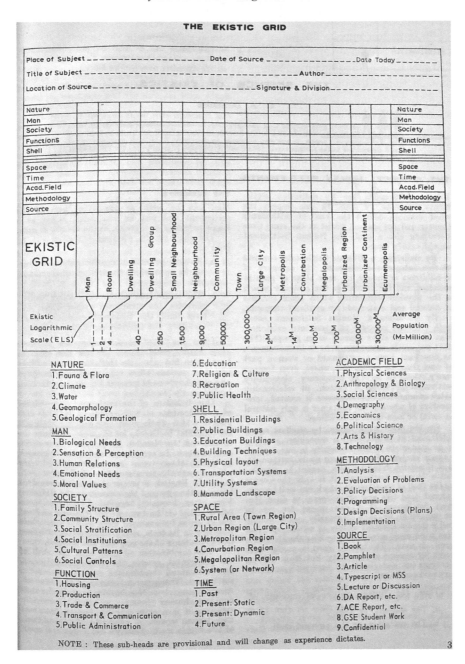

THE EKISTIC GRID

Place of Subject _ _ _ _ _ _ _ _ _ _ _ _ _ _ _ _ _ _ Date of Source _ _ _ _ _ _ _ _ _ _ _ _ _ Date Today _ _ _ _ _ _ _

Title of Subject _ Author _ _ _ _ _ _ _ _ _ _ _ _ _ _ _ _ _

Location of Source _ Signature & Division _ _ _ _ _ _ _ _ _ _ _ _ _ _ _ _

NATURE	6.Education	ACADEMIC FIELD
1.Fauna & Flora	7.Religion & Culture	1.Physical Sciences
2.Climate	8.Recreation	2.Anthropology & Biology
3.Water	9.Public Health	3.Social Sciences
4.Geomorphology		4.Demography
5.Geological Formation	SHELL	5.Economics
	1.Residential Buildings	6.Political Science
MAN	2.Public Buildings	7.Arts & History
1.Biological Needs	3.Education Buildings	8.Technology
2.Sensation & Perception	4.Building Techniques	
3.Human Relations	5.Physical layout	METHODOLOGY
4.Emotional Needs	6.Transportation Systems	1.Analysis
5.Moral Values	7.Utility Systems	2.Evaluation of Problems
	8.Manmade Landscape	3.Policy Decisions
SOCIETY		4.Programming
1.Family Structure	SPACE	5.Design Decisions (Plans)
2.Community Structure	1.Rural Area (Town Region)	6.Implementation
3.Social Stratification	2.Urban Region (Large City)	
4.Social Institutions	3.Metropolitan Region	SOURCE
5.Cultural Patterns	4.Conurbation Region	1.Book
6.Social Controls	5.Megalopolitan Region	2.Pamphlet
FUNCTION	6.System (or Network)	3.Article
1.Housing		4.Typescript or MSS
2.Production	TIME	5.Lecture or Discussion
3.Trade & Commerce	1.Past	6.DA Report, etc.
4.Transport & Communication	2.Present: Static	7.ACE Report, etc.
5.Public Administration	3.Present: Dynamic	8.GSE Student Work
	4.Future	9.Confidential

NOTE : These sub-heads are provisional and will change as experience dictates.

Figure 15.8   The Ekistic Grid features a logarithmic scale of settlements

In order to complete the job of designing the documentation center, Tyrwhitt extended her leave of absence from Harvard for another year. She had learned that preparing information for machine handling was more difficult than was once thought. Before automating an information system there was a need first to organize a monthly master index of all the relevant knowledge that had been generated, and then to abstract all useful documents. Tyrwhitt put an experimental system for indexing the content of *published* material of direct relevance to DA and ACE personnel— using a UNIVAC 1004 computer –into operation in January 1966. Tyrwhitt reported on the progress of this experiment at the Sixth Architectural Libraries Conference at RIBA in March 1966. By then she was ready for a new challenge.

Re-turning East

A new path opened up for Tyrwhitt when she went to Japan for the first time, shortly after the RIBA conference. The impetus for the trip was to visit her niece, Catharine Huws, a geographer who had met and married architect, Koichi Nagashima, when

RESEARCH GROUP ON DOCUMENTATION METHODS IN THE FIELD OF EKISTICS
Left to right: Mr Harold Overton, Prof. Gordon Bell, Prof. Gwen Bell (hidden), Mrs Newling (guest), Mr Th. Theodoridis, Mrs Vivian Sessions, Mr Constantinos Kakissopoulos, Mr Dan Fink, Prof. Jaqueline Tyrwhitt, Mrs M. Avronidakis (reporter)

tation centres, if they are related to a specialized institution whose experts coopèrate on the prepara-tion of such reports.   *provide current awareness o[field of interest; (b) as a tool b can be retrieved via its cla*

Figure 15.9   Tyrwhitt convened a meeting of experts on computer science, documentation and information management to discuss alternative information processing procedures

they were both working for Doxiadis. Nagashima spent a year at ACE after graduating in May 1964 from Harvard GSD, where he had been a student in the urban design studio Tyrwhitt co-taught with Maki. Koichi and Catharine were married in Athens in September 1965, held their reception at Sparoza, and promptly moved to Japan, where Koichi started work as a designer in Maki's Tokyo office. Tyrwhitt now had family ties to the close knit group of Japanese architects and planners whose urban design methods she found particularly promising.

Inspired by the Metabolists' ideas she had explored with Maki in the Harvard studio, she decided to undertake a study of the villages on the Greek island of Chios. Nagashima, along with many other ACE students helped Tyrwhitt make measured drawings of the villages, said to be established by the Greeks just after the Trojan war. Tyrwhitt considered these ancient settlements as prototypes for dense, human scale urban forms. They "are exceptionally interesting to modern eyes for their compact but flexible urban megastructure, for their separation of different functions on different levels and for their single community centers" (1966b: 475). Both Maki and Tange were Tyrwhitt's guests at Sparoza during the visitor season in 1965, and undoubtedly encouraged this line of investigation.

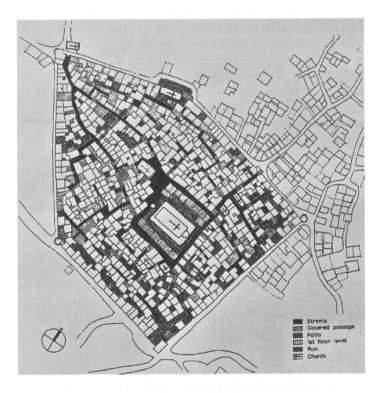

Figure 15.10 Tyrwhitt studied ancient villages on the Greek island of Chios as models for a compact urban megastructure

After Catharine and Koichi's wedding, Tyrwhitt took some time off from ACE to settle into her new house, to write the Chios article, and also to translate Giedion's new material for the fifth edition of *STA*. This edition captured how their symbiotic creative collaboration—their shared body of ideas—influenced Giedion's writings in English. Giedion had included a new chapter, "Changing Notions of the City," in this edition, "in view of various present-day tendencies, such as those manifested in the works of Kenzo Tange and Fumihiko Maki … . In place of the rigid master plan proposed in the early years of the century a flexible 'master program' is now being put forward, one that allows for changes and that leaves open-ended possibilities for the future" (1967: v, 862).[2] This was the quality that Tyrwhitt admired in the villages of Chios, as well as in the architecture of Tange and Maki. "But then [as a] planner … I have a weakness for an extensible system rather than a self-contained unit," she told Giedion. [SG.121] She invited Giedion to visit Japan with her, and he very much wanted to do that, but he was immersed in his next book on his theory of the three space conceptions of architecture.

Tyrwhitt timed her first visit to Japan to coincide with the 28th World Congress of IFHP in Tokyo in May 1966. This was "a wonderful occasion to become acquainted with a country that—at least on the surface—has somehow managed to adopt the technical know-how and engineering skills of the west without discrediting its high oriental civilization … . The appearance of this happy combination gives one a great confidence in the future." [E.22] Those words could also describe her pleasure at seeing Catharine and Koichi at their home in Zushi, a seaside town outside of Tokyo. Tyrwhitt was eager to spend more time with them, especially when their children were born; she jumped at the chance when sociologist Eichi Isomura invited her to propose a short research project to undertake at the Japan Centre for Area Development Research (JCAR), a group he had formed with support from Ford Foundation and the UN. [T.103] Tyrwhitt knew Isomura through his participation in Delos 1 and 2, but they had first met in 1959 when he spent a term as a fellow at the Harvard MIT Joint Center. Tyrwhitt proposed a study of pedestrian areas in Tokyo to Isomura, which was funded.

The Tokyo study involved surveying and mapping pedestrian behavior in two districts where narrow roads discouraged through traffic. She saw those districts—like the villages on Chios—as a form of urban mega-structure, "pedestrian islands," that could be used as building blocks in a large urban complex; they were human-scaled communities where the basic daily life needs could be met within safe and convenient walking distances. Tyrwhitt saw her contribution as lending support to this approach to the design of urban residential areas, and providing "a few useful suggestions and dimensions." [E.32] Like her survey of Ryerson students in 1955, Tyrwhitt's study of pedestrian behavior in Tokyo pioneered a path now well

---

2   IT is interesting to note that Tyrwhitt republished this article in 1968 in *Architect's Year Book*, Vol. 12, which also includes her edited version of Geddes' "The Valley Section," as well as an article by Doxiadis on "Linear Cities," which she probably edited, and which probably draws on her historical review of "shapes of cities that can grow."

trodden: research to standardize the "walk-able neighborhood" as a planning unit.[3] What is more significant is the way in which she proceeded holistically, conducting field research to develop new standards for practice at the same time that she was designing and helping to implement new programs for professional education and building relationships to promote interchange among schools of planning—and disseminating information about this worldwide through *EKISTICS*.

Returning to Harvard in September 1966, after her two-year leave of absence, Tyrwhitt seized the opportunity to capitalize on being in Tokyo to strengthen academic exchanges with planning schools in East Asia. Learning that a resolution was passed at an October 1966 UN Seminar on Planning for Urban and Regional Development to survey all of the existing and proposed planning schools in the ECAFE region to advise on the best use of international funds in that area, Tyrwhitt used her contacts at Ford Foundation and the UN to land two additional jobs: to advise on setting up a new planning school in Singapore; and to propose a system of reciprocal exchanges among schools of planning both within the South East Asian region and between those schools and Harvard. She spent much of fall 1967 traveling in SE Asia, visiting Japan, Singapore, the Philippines, India, and Indonesia. [T.104, 105]

On her way back to Athens Tyrwhitt spent a couple of days working with Giedion in Zurich in January 1968. Shortly after that visit Giedion, who had not been well for some time, cracked his pelvis. Preoccupied with Giedion's deteriorating condition, Tyrwhitt returned twice to visit him in the hospital. He never recovered, passing away at age 80, in April 1968. They had been looking forward to being together at Harvard in the fall. Giedion wrote her the day before he died: "I would be glad if we could be living the fall in one of Sert's student houses!" [SG.121] Their complicated, two decade-long relationship ended, she completed the translation and guided the production of the English version of his last book, *Architecture and the Phenomena of Transition: The Three Space Conceptions in Architecture* (1971), as a labor of love. [T.106] Giedion's death significantly attenuated the ties that bound Tyrwhitt to Europe.

Her own family—which aside from her elderly aunts consisted of her sister Edrica and Edrica's children—was scattered; two of her nieces and their growing families were living in Asia. While she had happily settled down in Sparoza, Tyrwhitt welcomed any opportunity to visit them. In early 1969 Maki informed her that he had been talking to Nagashima and others about collaborating to pursue large scale urban projects in South Asia and "we all have agreed that we need your advice on this matter and active participation in the future if possible." Tyrwhitt responded: "Of course I would be absolutely delighted to help in any way possible—not necessarily paid; after all I am a member of the family." [T.107, 108] By then, the new planning program at the University of Singapore had been

---

3   The concept of "twenty minute neighborhoods" is incorporated as a standard building block of the Portland Plan, a strategy for future development adopted by the City of Portland in April 2012. Available at: http://www.portlandonline.com/portlandplan/index.cfm?c=56527 (acessed June 18, 2012).

approved along the lines she had recommended—to focus on urban design—and Nagashima had been hired as the first instructor. Tyrwhitt arranged consulting assignments in 1970, 1974, and 1981, mainly under Ford Foundation auspices, to Singapore, Hong Kong, Bangkok, and Tokyo, in part to underwrite trips to visit her growing Asian extended family network. [T.109]

Tyrwhitt also used her position as *EKISTICS* editor to focus attention on innovative urban forms and practices emerging in the fast growing cities of Asia, particularly the Tokaido megalopolis that stretched from Tokyo to Nagoya. In *Human Identity and the Urban Environment*, a compilation of articles from *EKISTICS* that was intended to serve as a reader on the "ekistical" approach, Tyrwhitt and her co-editor, Gwen Bell, featured a case study of Tokaido "to illustrate the process of arriving at synthesis in a single megalopolis." They argued: "With its ancient culture and singularly rapid technological development, the Tokaido megalopolis provides a unique bridge between the developed and developing worlds, rural and urban economies, and the Oriental and Occidental ways of life. In several respects it may be symbolic of a more distant urban future of ecumenopolis, when all parts of the globe will become linked to a single interlocking urban system, within which each individual can achieve his own identity" (1972: 36).

Another way of describing the urban evolution Tyrwhitt saw in Tokyo is the creation of order out of chaos, one of Doxiadis's maxims. Delian Richard L. Meier—who also taught at ACE 1962–1963—reported in *EKISTICS* in May 1967 that he discovered the "Foundation for a New Urbanism" in the dynamic chaos of Japanese cities, particularly Tokyo, a pattern that expedited the creative interaction of "institutions, public, private, cooperative and hybrid." [E.33] Architect and mathematician Christopher Alexander—whom Tyrwhitt had first met and encouraged as a student at GSD—provided the theoretical underpinning for this observation in his influential essay "A City Is Not a Tree," first published in 1965 and abstracted in *EKISTICS* in June 1967; he argued that urban structure is more like a "semi-lattice" that contains "overlap, ambiguity, multiplicity of aspect" than a simple, hierarchical "tree, which as a mental image "is comparable to the compulsive desire for neatness and order." [E.34] In the August 1969 *EKISTICS* Tyrwhitt included an article by J.M. Richards suggesting that Western architects and planners could learn a lesson from "the dynamic quality that has been a by-product of the anarchical growth and pop-art vitality of cities like Japan's" on "how to endow our cities with the same sense of popular participation without plunging … into functional chaos." [E.35] Tyrwhitt offered readers a method for such participatory planning in the next article, based on a paper Alexander gave in Tokyo in 1968, that presented an early version of his "pattern language." [E.36] Alexander found in Zen-inspired Japanese architectural traditions an exemplar of the "organic structure" of built form that resulted when people share a common, living, pattern language to make buildings and towns. John Friedmann made another connection with Zen ideas in the December 1969 *EKISTICS*, beginning with Lao Tzu, "All things will go through their own transformations" and concluding, "Innovative planners must learn to practice … the Tao of Planning." [E.37]

Thinking Globally

Tyrwhitt retired from Harvard in 1969, and began to work for Doxiadis full time. She hardly settled down, since she continued to travel abroad frequently for consulting and teaching assignments. The annual Delos Symposia structured her intellectual work for the next several years, and she helped structure the symposia as the only person who attended all of the conferences aside from Doxiadis and his wife, and Buckminster Fuller. Her job as General Secretary became more difficult as the number of observers on board the boat increased, a range of public programs were organized in Athens related to the symposion, and there was a need to keep more complete records of the discussions. "Her editorial work on conferences was like a searchlight penetrating darkening mist, piercing through ambiguities, repetitions and follies to some insight, some summary, some just-right idea that would otherwise never had come into sight," recalled legal scholar and Delian Earl Finbar Murphy. "In order to render this service she had to grasp the whole purpose of the conference, its potential significance." Finbar, a past president of WSE added: "Sometimes she had to do her editing work while chairing the meeting. But even when she was not in the chair formally and officially her view was that of the permanent, though necessarily unacknowledged chairman." [E.38]

By July 1968, in part due to the reverberations of the exchange of ideas and new intellectual friendships formed among the intellectual elite who gathered on Doxiadis' "floating think tank" and related meetings, the UN resolved to convene a Conference on Problems of the Human Environment in 1972 in Stockholm. With the UN conference in mind, and cognizant of the impending conclusion of the Ford Foundation grant that supported the symposia, Doxiadis decided to make Delos 10 in July 1972 the final one. Since Tyrwhitt planned on retiring as editor of *EKISTICS* in September 1972, she used her final years of editorial authority to underscore the core value of viewing the problems of the human environment through an ecological lens. This view of the earth as a whole was stunningly captured for many people by the Earthrise photograph taken from Apollo 8 in December 1968, but it was a view that Tyrwhitt had long held.

She devoted the entire March 1969 issue of *EKISTICS* to Ecosystems: Man and Nature—a first for the journal. "Not at all long ago, it was difficult to find any material at all dealing with the interaction of man and the natural world," Tyrwhitt observed in her editorial: "Now people from almost every discipline are up in arms about the grave consequences of man's abuses of the—formerly seemingly inexhaustible—elements of air and water." [E.39]

Tyrwhitt continued to dedicate entire issues of *EKISTICS* to the theme of Ecosystems: Man and Nature in 1970 and 1971. She wrote in a Foreword in April 1970, "There are indications that words are leading to action," noting that the stated goal of the upcoming UN conference "is to bring about the adoption of action-oriented programs by public authorities at the local, national, regional and international levels to deal with the problems of planning, management and control of the human environment for economic and social developments." [E.40] This issue began with several articles that advocated a systems approach to environmental change, notably as proposed by Eugene Odum, the pioneer of ecosystem ecology. Introducing the 1971 issue on Man and Nature, she wrote about advances in knowledge since 1969, the institution of some controls, expressed, for example, in the National Environmental Policy Act in the US, and "indications of greater international cooperation in environmental management." There was now widespread understanding of the intrinsic connections between the health of people and places, even if "the fuse of the environmental time bomb grows shorter and shorter." [E.41]

In preparation for the final Delos Symposion, Tyrwhitt used the April 1972 issue of *EKISTICS* to synthesize excerpts from papers and remarks made during the summer international discussions from 1963–1971. She proudly reported in her Foreword: "Altogether, the quotations reflect the views of 138 people from Australia, Brazil, Canada, Chile, Denmark, France, Germany, Ghana, Greece, Egypt, India, Japan, Nigeria, Pakistan, Poland, Rumania, South Africa, Sweden, Switzerland, Uganda, the United Kingdom, The United States and the Union of Soviet Socialist Republics. The speakers quoted are also drawn from a great variety of disciplines: historians, natural scientists, social scientists, mathematicians, architects, engineers, artists and politicians." [E.42] This issue stands as a testimonial to Tyrwhitt's role in capturing the essence of these discussions, while retaining the distinctive qualities of multiple voices and native tongues.

The April 1972 issue of *EKISTICS* amplified the Delian voices who contributed to *Only One Earth: The Care and Maintenance of a Small Planet*, an unofficial report commissioned by the UN. Co-authored in 1971 by Renee Dubos and British economist and writer Barbara Ward (Lady Jackson)—a central figure in the Delos symposia—it was intended to provide a conceptual framework for participants in the 1972 Stockholm Conference. Their text incorporated the international discussions among Delians that Tyrwhitt had recorded, edited and disseminated through *EKISTICS*. Both Dubos and Ward participated in Delos 10 in July 1972, which was focused on the theme of Action for Human Settlements. That conference contributed to the establishment of the UN Environmental Program (UNEP), as well as to a call for a larger conference on the subject; Canada hosted the first UN Conference on Human Settlements (Habitat I) in 1976.

When asked to compare the Delos Symposia to the CIAM congresses, Tyrwhitt observed: "Without the CIAM congresses, it is very doubtful if the more humane approaches to urban planning would have been developed so early. … Without the Delos Symposia, it is doubtful that the United Nations would have got world support for its conferences on the environment and human settlements" (1978: 131). Tyrwhitt's transnational, holistic efforts to formulate urban planning and design theories, build and sustain institutions to do planning and design research and education, and develop standards for professional practice were essential contributions to both.

**Figure 15.11 Drawings of participants at Delos VI, 1967, as seen by
Ruth Abrams, capture Tyrwhitt's central position in this
constellation of luminaries**

Chapter 16

Epilogue

Tyrwhitt's Utopia:
A Garden on a Greek Hillside

Jacky Tyrwhitt took her time looking for land on which to build her house in Greece. The site had to be within commuting distance of Athens and have easy access to an international airport; there had to be land suitable for cultivation; she wanted a view that changed with the seasons, and at least a glimpse of the sea; and she wanted there to be some established trees (Tyrwhitt 1998: 2). Her close friend and ACE colleague, John Papaioannou, helped "scour the Attica peninsula for her in search of suitable sites" Catharine Nagashima recalled.[1] When she found one with potential she would take her sleeping bag and spend the night there. In late summer 1962 she was convinced that she had found her place. Since she was leaving for the US the next day, she entrusted Papaioannou with negotiations to secure the property. This proved to be a difficult job which Papaioannou—who was devoted to Jacky—handled with finesse. [E.43] When she returned to Greece in May 1963, she saw the land he had been able to buy: a strip of land running to the top of a steep, rocky, wind-swept, south and east-facing hillside overlooking the plain of Mesoghia (now occupied by the new Athens airport), then a largely rural area about 45 minutes southwest of Athens by car. Her idea for the garden was to collect the local flora and simply mass it and group it and let it grow, but first she intended to spend a year observing and cataloguing indigenous plants; there was a lot to do before the job of gardening could begin.

Jacky's land, known as Sparoza, featured a panoramic view with a glimpse of the Aegean, and was completely undeveloped: no roads, electricity, water, or telephone. The practical considerations of building a house in this setting soon led her to the idea of building a community. She had to purchase additional property to secure a water supply, and she would have to control a larger swath of contiguous land to protect her view and immediate surroundings. Since she couldn't afford to do this by herself, she came up with a self-financing scheme: as she gradually bought adjacent plots of land, and assembled a viable building site, she sold it to finance the next purchase, and so on. But she sold each site with the proviso that the owner could build only a two-storey house, could not fence the land (allowing all owners to roam the property within a perimeter fence), and would share the cost of access roads and water.

---

1 Quotes from Catharine Nagashima are based on personal communication with the author June 2012.

This was a voluntary agreement, and Jacky only sold to like-minded people, artists, musicians, and intellectuals. Most of the buyers were Greek, but early owners included two of Jacky's closest friends: Delia (her sister-in-law) and Gwen Bell (a former student who succeeded Jacky as editor of *Ekistics* in 1972). Her idea was that the owners would form a "managing committee" to make decisions affecting the composite property [E.34]. In part, due to the sudden death of one of the key neighbors, this ideal never fully materialized. She was, however, sociable with her neighbors and discovered a real sense of community in Paiania, the nearby village of the family from whom she bought her land. In the summer of 1963, Jacky rented rooms there. Catharine, who lived there with her aunt, recalled: "Such was the hospitality of the villagers that we were frequently invited to their homes for our evening meal." They made enduring friendships that summer. The villagers helped the families at Sparoza with harvests of olives, grapes, and other crops, and in turn the Sparoza households shared portions of their oil, wine, and produce. Slowly, Jacky became "part of the existing local community, while at the same time enhancing it with what she would create at Sparoza."

As Jacky gradually acquired land, she also began restoring the landscape of Sparoza, which was nearly barren, except for a small olive grove. The soil was poor and thin. One of Jacky's goals was to become self-sufficient in vegetables. Early on she built raised beds to start a kitchen garden, and devised an overall plan to plant indigenous trees, including olive, lemon and fig trees, as well as others for shade and shape. This was not easy to do. Papaioannou supervised the planting, which involved using dynamite to blast holes in the bedrock. By the time she moved into her house, she had installed around 300 trees, and most of them survived. Just as trees were seen as symbols of the optimism that inspired post-war reconstruction in Britain, the trees Jacky planted at Sparoza can be seen as symbols of the hope she shared with her fellow Delians for achieving a sustainable global future through local action. Significantly, Jacky was creating Sparoza at the same time she was joining the Delians in organizing as the World Society for Ekistics.

In 1964, while her house was being built, Jacky spent a couple of days a week in the herbarium of the newly opened Goulandris Natural History Museum. She methodically documented each species in the herbarium's collection, making drawings and taking notes on habitat, in order to decide what to plant in her garden and how. Jacky's systematic approach both helped her develop an intimate acquaintance with her new locality and region, and signaled her concern about the deterioration of the Greek landscape, and the absence of local flora where it should be celebrated (that is, at resorts and around civic buildings). At that time there was little information available in English on Mediterranean gardening; the herbarium was her best available resource. Local nurseries carried few native species, so she traveled all over Greece collecting seeds, bulbs, and cuttings of wild plants, and worked hard to propagate them at Sparoza. Once established, those plants needed low maintenance and little water; she helped pioneer ecological landscaping in Greece.

Jacky also wanted her garden to look good in the summer, when visitors arrived, but summer was when most native plants would be dead or withered in

Joqueline Tyrwhitt Sparoza Christmas 1965

**Figure 16.1   Jacky's annual seasons greetings card featured an image of
Sparoza and an invitation to visit**

the dry heat. This meant she had to introduce non-native plants that would thrive in
the Mediterranean climate and in Sparoza's alkaline soil. She experimented with
many species, ordering seeds from the Royal Horticultural Society and collecting
plants from Botanical Gardens throughout the world, and acclimatizing them.
And she built a series of terraces where she improved and enriched the soil (Gay
2006). This painstaking process provides an apt metaphor for Jacky's approach
to international technical assistance: supplementing local material with imports
suitable for adaptation to particular local circumstances, thereby adding to the
vitality of the local ecosystem and seeding evolutionary change.

Jacky's garden promoted international exchange in a direct as well as a
metaphorical manner. After struggling for years to establish the garden with the
assistance of retired local workmen, when she reached age 70, Jacky realized
she needed more help. That's when she began retaining a series of young
horticulturalists that would live in an annex to her house for a year, and learn
about Greek flora while working with her in the garden. The system worked so
well that Jacky institutionalized it with the Royal Edinburgh Botanical Garden,
which agreed to select a suitable person for her from among their graduates. This
enduring transnational network of exchange magnified the impact of her mentoring
and the educational value of both her wild and cultivated gardens.

Sparoza became a node in a global network, as numerous visitors were attracted
there by Jacky's forceful personality, her generous hospitality, and her magnificent
house and garden. Jacky opened her home not only to friends and relations but also
to friends of friends and children of friends, former students and friends of theirs,

students who exchanged work on the property for seminars, dignitaries visiting Doxiadis or attending meetings at ACE, and Delians in town for the symposion or related events. Sparoza was both an informal annex of ACE and the core of Jacky's own constellation. Jacky found such entertaining exhilarating; she built a geodesic dome designed by Buckminster Fuller to create additional event space. The gatherings Jacky hosted created a venue for the kind of casual exchange and voluntary initiatives that she believed played an important role in developing international cooperation.

With the help of her young horticultural assistants, Jacky's garden thrived, despite periodic setbacks due to the vagaries of weather. After one storm she wrote in her diary: "It's one of the frustrations of gardening that one is so often having to start all over again." Jacky's ability to cultivate Sparoza as a garden, as an educational site, as a center for intellectual exchange, and as a community based on cooperation, is a testimony to her resiliency and her energy. But it also speaks to the long and mutually influential friendships she nurtured. Notably, in realizing her goal of turning Sparoza into a botanic preserve, Jacky received important encouragement from Gwen Bell, whom she considered "even closer than family." In 1979, Bell donated nearly half her property at Sparoza to Jacky's "botanic project." This gift provided the impetus Jacky needed to write her garden book, to inspire others to replicate her model in places with similarly harsh conditions, and to bequeath her house and garden to the Goulandris Natural History Museum so that it would benefit future generations. The impression that Jacky made as a mentor on younger people—Gwen among many others—is the reason her influence has reached beyond her era, leaving an indelible imprint. Her garden at Sparoza, a living work of art, survives as a lasting tribute to the intellectual, professional, and personal contributions she made throughout her remarkable life.

Mary Jaqueline Tyrwhitt

Making a Garden on a Greek Hillside

Figure 16.2 Jacky's book was published thanks to the Mediterranean Garden Society, which makes its home at Sparoza, and nurtures her garden for future generations

Bibliography

Archives and Primary Sources

References to material from archival or primary sources are cited in the text in brackets; abbreviations refer to the archive and numbers refer to the list of references from that archive. In the case of Tyrwhitt's diaries, the numbers refer to a date [month/day/year].

Abbreviation	Archive or Primary Source
[month/day/year]	Tyrwhitt personal archive: diaries and travel journals, to be transferred to the Tyrwhitt collection in the RIBA archive (see below).
SG	Tyrwhitt personal archive: Giedion Correspondence, to be transferred to the Tyrwhitt collection in the RIBA archive (see below).
T	Tyrwhitt papers: Royal Institute of British Architects (RIBA) Architectural Library Drawings & Archives Collection, London.
Ty	Tyrwhitt papers in the Sir Patrick Geddes Collection, Strathclyde University Archives, Andersonian Library, Glasgow.
G	Patrick Geddes Centre for Planning Studies, Edinburgh University Library Special Collections, Edinburgh.
B	Catherine Bauer Wurster papers 1931–64, Bancroft Library, University of California Berkeley.
L	CIAM archive, Jose Luis Sert archive and Special Collections: Frances Loeb Library, Harvard Graduate School of Design, Cambridge.
F	Ford Foundation Archive, New York.
D	Constantinos A. Doxiadis Archive, Benaki Museum, Athens.
E	*EKISTICS*, the journal is considered here as a primary source.

Tyrwhitt Personal Archive: Giedion Correspondence		
1. SG to JT. 1/24/48.	20. JT to SG. 1/9/49.	39. SG to JT. 12/20/51.
2. SG to JT. 9/26/47.	21. JT to SG. 1/16/49.	40. SG to JT. 8/7/50.
3. SG to JT. 9/29/47.	22. I. Gropius to JT. 2/3/49.	41. SG to JT. 12/2051.
4. SG to JT. 10/27/47.	23. JT to SG. 7/7/29.	42. SG to JT. 7/31/51.
5. SG to JT. 11/7/47.	24. SG to JT. 8/9/49.	43. SG to JT. 11/11/51.

6. SG to JT. 1/7/48 [47].

7. SG to JT. 1/20/48.

8. SG to JT. 11/5/47.

9. SG to JT. 2/4/48.

10. SG to JT. 2/7/48.

11. SG to JT. 3/31/48.

12. SG to JT. 5/28/48.

13. SG to JT. 4/25/48.

14. SG to JT. 11/5/48.

15. SG to JT. 12/7/48.

16. SG to JT. 1/4/49.

17. SG to JT. 1/18/49 [48].

18. SG to JT. 12/30/48.

19. SG to JT. 1/24/49.

25. JT diary 9/5/49.

26. JT to SG. 10/14/49.

27. SG to JT. 10/17/49.

28. JT to SG. 10/26/49.

29. SG to W. Coates 11/25/49.

30. JT to SG. 1/6/50.

31. JT to SG. 1/11/50.

32. JT to SG. 11/28/49.

33. JT to SG. 12/25/49.

34. JT to SG. 1/14/50.

35. JT to SG. 1/20/50.

36. SG to JT. 3/22/50.

37. JT to SG. 3/21/50.

38. SG to JT. 4/11/50.

44. JT to SG. 1/6/52.

45. JT to SG. 1/11/53.

46. JT to SG. 2/18/53.

47. JT to SG. 1/18/53.

48. SG to JT. 1/25/53.

49. JT to SG. 10/24/52.

50. JT to SG. 12/19/52.

51. JT to SG. 3/8/53.

52. JT to SG. 3/30/53.

53. JT diary 6/3/53.

54. JT to SG. 6/20/53.

55. JT to SG. 8/12/53.

56. JT to SG. 9/6/53.

57. JT to SG. 12/6/53.

58. JT to SG. 11/22/53.

59. JT to SG. 11/7/53.

60. JT to SG. 10/18/53.

61. JT to SG. 12/30/53.

62. Sevagram Rept. 12/28/53.

63. JT to SG. 1/10/54.

64. JT to SG. 1/30/54.

65. JT to SG. 2/10/54.

66. SG to JT. 5/9/54.

67. JT to SG. 4/26?/54.

68. JT to SG. 6/20/54.

69. JT to SG. 5/16/54.

70. JT to SG. 6/1/54.

71. SG to JT. 5/7/54.

72. JT to SG. 8/13/54.

73. SG to JT. 9/6/54.

74. SG to JT. 10/3/74.

75. JT to SG. 8/27/54.

76. JT to SG. 10/17/54.

77. SG to JT. 5/15/55.

78. SG to JT. 1/4/55.

79. JT to SG. 1/23/55.

80. JT to SG. 3/19/55.

81. SG to JT. 1/26/55.

82. SG to JT. 4/2/55.

83. SG to JT. 5/15/55.

84. JT to SG. 11/11/55.

85. JT to SG. 3/20/55.

86. JT to SG. 4/6/55.

87. JT to SG. 5/3/55.

88. JT to SG. 3/31/55.

89. JT to SG. 5/19/55.

90. JT to SG. 2/15/55.

91. JT to SG. 6/15/55.

92. JT to SG. 6/1/55.

93. JT to SG. 6/8/55.

94. JT to SG. 6/17/56.

95. JT to SG. 5/28/56.

96. JT to SG. 5/7/56.

97. JT to SG. 9/3/56.

98. SG to JT. 9/13/56.

99. JT to SG. 9/17/56.

100. JT to SG. 1/20/57.

101. JT to SG. 7/6/57.

102. JT to SG. 11/22/56.

103. JT to SG. 7/19/57.

104. JT to SG. 10/17/57.

105. JT to SG. 11/11/57.

106. SG to JT. 10/10/58.

107. SG to JT. 10/24/58.

108. SG to JT. 8/23/59.

109. JT to SG. 1/19/60.

109. JT diary 9/19/53.

110. JT to SG. 3/1/60.

111. JT to SG. 1/19/60.

112. JT to SG. 2/18/60.

113. JT to SG. 7/19/58.

114. JT to SG. 3/16/63.

115. JT to SG. 5/2/63.

116. SG to JT. 6/27/64.

117. JT to Marg't 7/21/64.

118. JT to SG. 4/6/63.

119. JT to SG. 5/14/64.

120. JT to SG. 9/14/64.

121. JT to SG. 5/10/65.

Tyrwhitt papers: RIBA

1   Amesbury Farm Settlement. N.d. Experimental Cottages, Ministry of Agriculture and Fisheries. TyJ/66/15.

2   Bull, S. J. 1930. Letter To whom it may concern. August 15. TyJ/1/4

3   Tyrwhitt, J. 1953. School of Planning. March 3. TyJ/38/2.

4   Tyrwhitt, J. 1953. Text of interview for U. Toronto student magazine. March. TyJ/39/11.

5   The School of Planning and Research for Regional Development. 1952. A Short Account of its History; Aims and Objects and Proposals for its Future Development. May. London. TyJ/6/2.

6   Rowse, E. A. A. 1938. Letter to J. Tyrwhitt. January 24. TyJ/1/4.

7   Tyrwhitt, J. 1969. Letter to A. M. Vogt, Personal Note on S. Giedion. January 12. TyJ/60/2.

8   Tyrwhitt, J. 1939. Letter to L. Elmhirst. September 26. TyJ/1/6.

9   Tyrwhitt, J. 1939. Letter to E. A. A. Rowse. September 11. TyJ/1/7.

10   Rowse, E. A. A. 1939. Letter to J. Tyrwhitt. October 27. TyJ /1/7.

11   Coombs, G. T. 1939. Letter to J. Tyrwhitt. November 20. TyJ/1/5.

12   Storey, F. 1939. Letter to R. E. Marsden. December 2. TyJ/1/5.

13   Rowse, E. A. A. 1940. Letter to J. Tyrwhitt. March 20. TyJ/1/7.

14   Davies, F. 1940. Letter to J. Tyrwhitt. March 21. TyJ/1/7.

15   Tyrwhitt, J. 1940. Draft Letter to Mr. Young. July 8. TyJ/1/6.

16   Pott, A. 1941. Letter to J. Tyrwhitt. February 17. TyJ/1/6.

17   Tyrwhitt, J. 1945. Resume. March 2. TyJ/66/1.

18   Tyrwhitt, J. 1943. Tentative Survey of Local Food Customs in England and Wales. October 9. TyJ/9/2.

19   Hartland, M. 1945. Letter to J. Tyrwhitt. February 5. TyJ/46/1.

20   Tyrwhitt, J. 1948. Letter to A. Mayer. September 9. TyJ/47/6.

21   Tyrwhitt, J. 1946. Handwritten notes about journey to Paris to attend the *Congrès Technique Internationale*. September 16–21. TyJ/45/7.

22   Forrester, J. 1947. Letter to J. Tyrwhitt. June 19. TyJ/47/3

23   A Faculty of Interrelations. 1942. Paper presented at Middle Atlantic States Art Conference, second session. Philadelphia Museum of Art. TyJ/59/14.

24   Tyrwhitt, J. 1948. New Housing in New York City—First Impressions. March. TyJ/39/4.

25   Abrams, C. 1948. *Post-Home News Housing Adviser*. April 16. TyJ/39/3.

26   Tyrwhitt, J. 1950. Letter to B. Wills. January 23. TyJ/48/1.

27   Chronology and Details of her Career. 1953. April 3. TyJ/48/7.

28   The Size and Shape of Urban Communities. 1949. Lecture given at Vassar College Faculty Club. February 14. TyJ/38/2/1.

29   VII CIAM Congress. 1949. Report of Commission IV. July 29. TyJ/66/3.

30   Tyrwhitt, J. 1965. Le Corbusier and CIAM. *GSD Newsletter* 12(1). TyJ/28/3.

31   VII CIAM Congress. 1949. Report of Commission IV. July 29. TyJ/66/3.

32   Tyrwhitt, J. 1949. Letter to Gordon Stephenson. November 8. TyJ/47/4.

33   Tyrwhitt, J. 1949. Letter to J. Forrester. December 5. TyJ/48/1.
34   Tyrwhitt, J. 1951. Town Planning on Display. British Information Services. N.p. March 21. TyJ/14/13.
35   Giedion, S. 1951. Letter to J. Tyrwhitt. March 12. TyJ 59/17.
36   Tyrwhitt, J. 1951. Letter to S. Giedion. April 25. TyJ/59/17.
37   CIAM 8. 1951. Open Session Background of the Core. July 9. TyJ/45/6.
38   Tyrwhitt, J. 1951. Letter to S. Giedion. May 24. TyJ 59/17.
39   Giedion, S. 1951. Letter to J. Tyrwhitt. May 15. TyJ/59/17.
40   Tyrwhitt, J. 1951. The Constellation and the Core. Talk given to the Council for Planning Action in Hunt Hall, Cambridge. November 27. TyJ/39/3.
41   Tyrwhitt, J. 1952. Letter to E. A. A. Rowse. January 11. TyJ/48/2.
42   Rowse, E. A. A. 1952. Letter to J. Tyrwhitt. January 1. TyJ/48/2.
43   MARS Group. 1952. Minutes of Special General Meeting. June 5. London. TyJ/66/22.
44   Pepler, G. 1952. Letter to J. Tyrwhitt. February 1. TyJ/48/2.
45   Weissmann, E. 1952. Letter to Mr. Gurney. July 23. TyJ/48/2.
46   Gurney. 1952. Letter to J. Tyrwhitt. September 5. TyJ/48/2.
47   Tyrwhitt, J. 1952. Letter to C. Ross. December 2. TyJ/48/2.
48   Tyrwhitt, J. 1954. Introduction to Reports of the Seminar. *UN Regional Seminar on Housing and Community Improvement, New Delhi, India, 20 January–17 February*, Housing and Town and Country Planning Section, Dept. of Social Affairs. New York: United Nations. April 21. TyJ/28/6.
49   Tyrwhitt, J. 1953. Letter to E. Hinder. June 5. TyJ/31/9.
50   Tyrwhitt, J. 1954. History of UN Seminar on Housing and Community Improvement. May 3. TyJ/32/1.
51   Tyrwhitt, J. 1954. The Delhi Seminar on Housing and Planning for South East Asia. December 9. TyJ/29/1.
52   United Nations Technical Assistance Program. 1954. *UN Seminar on Housing and Community Improvement in Asia and the Far East, New Delhi, India 21 Jan–17 Feb 1954*. December 1. TyJ/28/4.
53   Tyrwhitt, J. 1953. Letter to E. Hinder. June 12. TyJ/31/9.
54   Program of the Regional Seminar on Housing and Community Improvement. N. d.
55   TyJ/28/5.
56   Tyrwhitt, J. 1953. Letter to E. Hinder. July 14. TyJ/31/9.
57   Tyrwhitt, J. 1953. Letter to E. Weissmann. August 1. TyJ/31/9.
58   Tyrwhitt, J. 1954. Circular Letter, Seminar on Housing and Community Improvement, Exhibition Grounds, New Delhi. February 12. TyJ/32/1.
59   Tyrwhitt, J. 1956. Creation of the Village Center. *Proceedings of the South East Asia Regional Conference in New Delhi, February 1-7 1954, organized by the International Federation for Housing &*

Town Planning, in conjunction with the UN Seminar. pp. 220–25, here 220. TyJ/29/2.

60 Hinder, E. 1954. Letter to J. Tyrwhitt. January 19. TyJ/31/9.

61 Weissmann, E. N. d. Closing Statement. *Housing and Town and Country Planning Section*, Department of Social Affairs, UN, United Nations Regional Seminar. TyJ/28/6.

62 Tyrwhitt, J. 1954. Princeton School of Architecture Seminar on Village Buildings for the Central Areas of India. May. TyJ/31/9.

63 Langford, G. 1953. Letter to J. Tyrwhitt. April 30. TyJ/18/2.

64 Tyrwhitt, J. 1953. Letter to Joe. June 11. TyJ/18/2

65 Lawson, M. 1954. Letter to J. Tyrwhitt. Apr 9. TyJ/18/5.

66 Tyrwhitt, J. 1953. Letter to M. McLuhan. August 30. TyJ/18/2.

67 Tyrwhitt, J. 1954. Letter to M. McLuhan. May 10. TyJ/18/2.

68 McLuhan, M. 1953. Letter to J. Tyrwhitt. October 29. TyJ/18/2.

69 McLuhan, M. 1953. Letter to J. Tyrwhitt. December 8. TyJ/18/2.

70 Tyrwhitt, J. 1954. Letter to M. McLuhan. May 16. TyJ/18/2.

71 Giedion, S. 1953. Some Words for the Japanese Edition of 'Space, Time & Architecture, revised version by Tyrwhitt. April. TyJ/59/3.

72 Tyrwhitt, J. 1954. Letter to J. L. Sert. November 4.

73 Sert, J. L. 1954. Letter to J. Tyrwhitt. November 29. TyJ/48/3.

74 Carver, H. 1954. Letter to J. Tyrwhitt. May 21. TyJ/18/2.

75 Tyrwhitt, J. 1954. Letter to T. Adamson. May 10. TyJ/18/2.

76 Isaac, R. 1955. Letter to J. Tyrwhitt. March 16. TyJ/48/3.

77 Daves, J. Letter to J. Tyrwhitt. TyJ/59/18.

78 Doxiadis, C. 1955. Letter to J. Tyrwhitt. April 8. TyJ/48/3.

79 Tyrwhitt, J. 1955. *Tropical Housing & Planning Monthly Bulletin.* Dec 13. TyJ/48/3.

80 Urban Design Roundtable. 1956. November 26-27. TyJ/26/3.

81 Lynch, K. 1957. Letter to J. Tyrwhitt. January 4. TyJ/48/6.

82 Tyrwhitt, J. 1957. Letter to K. Lynch. January 22. TyJ/48/6.

83 Sert, J. L. and J. Tyrwhitt. 1957. The Shape of Our Cities: A Series of Experimental Study-Discussion Programs on Urbanism. TyJ/23/6.

84 Tyrwhitt, J. 1960. Letter to J. L. Sert. March 17. TyJ/18/6.

85 Sert, J. L. 1960. Letter to J. Tyrwhitt. March 24. TyJ/18/6.

86 Meyerson, M. 1960. Memo to Jason Epstein Jul 27. TyJ/42/8.

87 Giedion, S. 1959. Letter to V. Gilman. March 10. TyJ/54/5.

88 Holliday, C. 1953. Letter to J. Tyrwhitt. November 13. TyJ/31/7.

89 Tyrwhitt, J. 1953. Letter to C. Holliday. December 4. TyJ 31/7.

90 Tyrwhitt, J. 1958. Letter to G. Druyor. April 29. TyJ/48/6.

91 Doebele, W. 1959. Letter to M. Meyerson. June 3. TyJ/19/3.

92 Tyrwhitt, J. 1960. Letter to J. L. Sert. March 17. TyJ/18/6.

93 Tyrwhitt, J. 1960. Final Report. July 28. TyJ/33/2.

94 Alonso, W. 1960. Letter to M. Meyerson. December 2. TyJ 33/5.

95 Meyerson, M. 1960. Letter to J. Tyrwhitt. July 8. TyJ/20/5.

96   Minutes of meeting with Prof. Tyrwhitt. 1960. Indonesia—Institute of Town and Regional Planning. October 13. TyJ/33/5.
97   Moehtadi, M. O. B. 1961. Memo to J. Tyrwhitt. Mar 30. Tyrwhitt draft reply. TyJ/33/5.
98   Conference for Future Faculty, Harvard University. 1961. Report. May 28–30. TyJ/33/1.
99   Wirioatmodjo, S. 1963. Letter to J. Tyrwhitt. July. TyJ/33/5.
100   Jackson, H. 1960. Letter to J. Tyrwhitt. Mar 2. TyJ/18/6.
101   Isaacs, R. 1960. Letter to J. Tyrwhitt. May 9. TyJ/18/6.
102   Meyerson, M. 1960. Letter to J. Tyrwhitt. July 8. TyJ/20/5.
103   Urban Design 6: Designing for Intercity Growth. 1962. TyJ/27/1.
104   Harvard Graduate School of Design. 1966. A Documentation Center on Ekistics. *Alumni Association Newsletter*, 12(2). TyJ/28/3.
105   Tyrwhitt, J. 1966. Report on visit to Japan May 4-23. TyJ/36/6/3.
106   Tyrwhitt, J. 1967. Urban Planning, a memorandum for the Singapore Polytechnic. February. TyJ/50/1.
107   Tyrwhitt, J. 1967. Letter to Prof. Sugianto. August. TyJ/60/2.
108   Tyrwhitt, J. 1968. Letter to D. Clemente. March 9. TyJ/60/1.
109   Maki, F. 1969. Letter to J. Tyrwhitt. February 6. TyJ/50/4.
110   Tyrwhitt, J. 1969. Letter to F. Maki. February 17. TyJ/50/4.
111   Tyrwhitt, J. 1970. Memorandum to George F. Gant. January. TyJ/37a/1-4.

Patrick Geddes Center for Planning Studies: Edinburgh University Library

1   Tyrwhitt, J. 1939. Letter to E. A. A. Rowse. October 21. PG/1.5.
2   Tyrwhitt, J. 1939. Letter to Davies. November 8. PG/1.5.
3   Tyrwhitt, J. 1940. Letter to E. A. A. Rowse. March 28. PG/1.5
4   Rowse, E. A. A. 1940. Letter to P. Cocke. March 18. PG/1.4.
5   Rowse, E. A. A. 1940. Letter to J. Tyrwhitt. March 21. PG/1.5.
6   Tyrwhitt, J. 1940. Letter to E. A. A. Rowse. March 28. PG/1.5.
7   Rowse, E. A. A. 1940. Letter to J. Tyrwhitt. April 2. PG/1.5.
8   Rowse, E. A. A. 1940. Letter to J. Tyrwhitt. April 13. PG/1.5.
9   Tyrwhitt, J. 1940. Letter to E. A. A. Rowse. April 15. PG/1.5.
10   Rowse, E. A. A. 1940. Letter to J. Tyrwhitt. April 20. PG/1.5.
11   Tyrwhitt, J. 1940. Letter to E. A. A. Rowse. April 25. PG/1.5.
12   Rowse, E. A. A. 1940. Letter to J. Tyrwhitt. May 3. PG/1.5.
13   Tyrwhitt, J. 1940. Letter to E. A. A. Rowse. June 14. PG/1.5.
14   Tyrwhitt, J. 1940. Letter to E. A. A. Rowse. July 1. PG/1.5.
15   Rowse, E. A. A. 1940. Letter to J. Tyrwhitt. July 5. PG/1.5.
16   Rowse, E. A. A. 1940. Letter to J. Tyrwhitt. July 19. PG/1.5.
17   Tyrwhitt, J. 1940. Letter to E. A. A. Rowse. July 23. PG/1.5.
18   Forestry Commission. 1940. Letter to E. A. A. Rowse. August 13. PG/1.5.
19   Tyrwhitt, J. 1940. Letter to E. A. A. Rowse. August 4. PG/1/5.
20   Rowse, E. A. A. 1940. Letter to Forestry Commission. August 10. PG/1.5.

21   Rowse, E. A. A. 1940. Letter to J. Tyrwhitt. August 10. PG/1.5.
22   Tyrwhitt, J. 1940. Letter to E. A. A. Rowse. August 16. PG/1.5.
23   Rowse, E. A. A. 1940. Letter to J. Tyrwhitt. August 25. PG/1.5.
24   Tyrwhitt, J. 1940. Letter to E. A. A. Rowse. September 3. PG/1.5.
25   Tyrwhitt, J. 1940. Letter to E. A. A. Rowse. September 7. PG/1/5.
26   Rowse, E. A. A. 1940. Letter to P. Scott. August 11. PG/1.4.
27   Rowse, E. A. A. 1940. Letter to M. Radford. September 12. PG/1.4.
28   Rowse, E. A. A. 1940. Letter to J. Tyrwhitt. September 26. PG/1.5.
29   Scott, P. 1940. Letter to E. A. A. Rowse. November 6. PG/1.4.
30   Rowse, E. A. A. 1940. Letter to J. Tyrwhitt. Nov 19. PG/1.5.
31   Geddes, A. 1947. Letter to J. Tyrwhitt. October 6. PG/17.

Catherine Bauer Wurster Papers: Bancroft Library

1   Schoendorff, E. 1950. Letter to C. Bauer. January 7. Box 10.
2   Tyrwhitt, J. 1946. Letter to C. Bauer. July 11. Box 10.
3   Bauer, C. 1946. Letter to J. Tyrwhitt. August 19. Box 2
4   Tyrwhitt, J. 1946. Letter to C. Bauer. October 16. Box 10.
5   Tyrwhitt, J. 1947. Letter to C. Bauer. November 11. Box 10.
6   Bauer, C. 1947. Letter to J. Tyrwhitt. November 26. Box 2
7   Tyrwhitt, J. 1949. Letter to C. Bauer. November 7. Box 10.
8   Tyrwhitt, J. 1949. Letter to C. Bauer. March 31. Box 10.
9   Tyrwhitt, J. 1949. Letter to C. Bauer. May 4. Box 10.
10   Bauer, C. 1949. Letter to J. Tyrwhitt. May 25. Box 3
11   Tyrwhitt, J. 1950. Letter to C. Bauer. March 7. Box 10.
12   Tyrwhitt, J. 1950. Letters to C. Bauer. March 7; April 7. Box 10.
13   Bauer, C. 1952. Notes on my UN Assignment. April 1. Box 4.

CIAM, Sert and Special Collections: Frances Loeb Library

1   Giedion, S. 1950. Letter to J. L. Sert. January 12. CIAM C007.
2   Giedion, S. 1950. Letter to J. L. Sert. November 21. CIAM C007.
3   Sert, J. L. 1950. Letter to J. Tyrwhitt. July 18. CIAM C007.
4   Tyrwhitt, J. 1950. Letter to J. L. Sert. August 8. CIAM C007.
5   Sert, J. L. 1950. Letter to S. Giedion. August 15. CIAM C007.
6   Sert, J. L. 1950. Letter to J. Tyrwhitt. November 9. CIAM C007.
7   Giedion, S. 1950. Letter to J. L. Sert. November 21. CIAM C007.
8   Tyrwhitt, J. 1950. Letter to J. L. Sert. December 13. CIAM C007.
9   Sert, J. L. 1951. Letter to J. Tyrwhitt. August 1. CIAM C009.
10   Sert, J. L. 1951. Letter to J. Tyrwhitt. August 30. CIAM C009.
11   Tyrwhitt, J. 1951. Programme of Lectures on Principles of Town Planning. University of Toronto, Fall Term 1951. CIAM E005.
12   Tyrwhitt, J. 1951. Letter to J. L. Sert. October 5. CIAM C010.
13   Sert, J. L. 1952. Letter to S. Giedion. June 12. CIAM C011.

14   Sert, J. L. 1952. Letter to J. Tyrwhitt. April 24. CIAM C011.
15   CIAM. 1952. Circular letter. May 14. CIAM C011.
16   Tyrwhitt, J. 1952. Letter to J. L. Sert. May 13. CIAM C011.
17   Tyrwhitt, J. 1952. Letter to J. L. Sert. June 8. CIAM C011.
18   Sert, J. L. 1952. Letter to S. Giedion. June 12. CIAM C011.
19   Tyrwhitt, J. 1952. Letter to J. L. Sert. August 3. CIAM C012.
20   Sert, J. L. 1952. Letter to J. Tyrwhitt. November 26. CIAM C012.
21   Tyrwhitt, J. 1952. Letter to J. L. Sert. December 2. CIAM C012.
22   Tyrwhitt, J. 1953. Letter to J. L. Sert. April 13. CIAM C013.
23   Kayanan, A. C. 1951. Memo to E. Weissmann: Notes on the ASPO
        National Planning Conference in Pittsburgh. October 25. CIAM C010.
24   Sert, J. L. 1953. Letter to J. Tyrwhitt. April 20. CIAM C013.
25   Short Report of Meeting in London. 1954. July 29. CIAM C015.
26   CIAM. 1955. Meeting of Delegates at La Serraz. Dec. 15. CIAM C017.
27   Sert, J. L. 1956. Introduction to the Urban Design Conference—
        Graduate School of Design, Harvard University. April 9. SERT D092.
28   Statements for Discussion Harvard Graduate School of Design: 2nd
        Urban Design Conference. 1957. April 12-13. SERT D093.
29   Second Urban Design Conference Graduate School of Design
        Harvard University. 1957. Introductory Notes. April. SERT D093.
30   Tyrwhitt, J. and J. L. Sert. 1958. Shape of American Cities. June.
        SERT C003-b.
31   Urban Design 3. 1959. General Report of Proceedings. April.
        SPECIAL HT107.U712x 1959
32   Urban Design 8. 1964. Concluding Meeting. May 2. SPECIAL
        HT107.U712x 1964

Ford Foundation Archive: McLuhan and Doxiadis

1   Changing Patterns of Language and Behavior and the New Media
        of Communication. 1953. University of Toronto Proposal to the
        Ford Foundation. Received March 23. Grant 53–70 Section 1.
2   News from the University of Toronto. 1953. May. Grant 53–70 Section 1.
3   McLuhan, M. 1955. Inter-Disciplinary Seminar in Culture and
        Communications at Toronto University 1953–1955. Received Aug
        5. Grant 53–70 Section 3.
4   Winnick, L. 1989. The Athens Center of Ekistics: The Urban World
        According to Doxiadis. Ford Foundation Urban History, Chapter
        VII, Part 3. May. Reports 012158.

Constantinos A. Doxiadis Archive

1   Doxiadis, C. 1956. Letter to J. Tyrwhitt. November 6. File 19248.
2   Doxiadis, C. 1957. Letter to J. Tyrwhitt. March 17. File 19248.

3   Doxiadis, C. 1963. Thoughts on the Delos Symposion. July 19. Doxiadis Associates Memo S–D 6504. File 118931.

EKISTICS Citation List

1   Tyrwhitt, J. 1985. Early career. *EKISTICS,* 52(314/315), 404.
2   Frankl, W. 1985. The Early Years—London. *EKISTICS,* 52(314/315), 421–2.
3   Rosenberg, G. 1985. The Early Years—London. *EKISTICS,* 52(314/315), 418–19.
4   Lock, M. The Early Years—London. *EKISTICS,* 52(314/315), 420–1.
5   Sekler, E. 1985. The Early Years—London. *EKISTICS,* 52(314/315), 422–3.
6   Johnson-Marshall, P. 1985. The Early Years—London. *EKISTICS,* 52(314/315), 416–18.
7   Carlson, E. 1985. One of the first—The world her professional habitat. *EKISTICS,* 52 (314/315), 489–92.
8   Bosman, J. 1985. My association with CIAM gave me a new perspective. *EKISTICS,* 52(314/315), 478–86.
9   Papaioannou, P. 1985. A short history of EKISTICS. *EKISTICS,* 52(314/315), 454.
10   Bauer, C. 1956. Economic development and urban living conditions: an argument for regional planning to guide community growth. Rough draft 1955, prepared for the Housing and Town and Country Planning Section of the United Nations. *Tropical Housing & Planning Monthly Bulletin,* 1(5), 1–4.
11   United Nations. Economic and Social Council. Social Commission. 1957. Report to the Economic and Social Council on the eleventh session of the Commission held in New York from 6 May 1957 to 25 May 1957. *Tropical Housing & Planning Monthly Bulletin,* 3(23), 310–13.
12   Editorial. 1957. *EKISTICS,* 4(25), iii.
13   Editorial. 1959. *EKISTICS,* 7(42), 291–2.
14   Editorial, 1959. *EKISTICS,* 7(44), 444.
15   Editorial. 1959. *EKISTICS,* 8(46), 79.
16   United Nations Seminar on Housing and Community Improvement. Seminar Working Party. 1957. Part I. United Nations program in housing building and planning International activities in New Delhi, India, January–March, 1954, Recent international activities in Asia and the Far East. *EKISTICS,* 3(18), 65.
17   Tyrwhitt, J. 1960. Preliminary Research Into Uses of House Plots By New Regional Planning School in Indonesia. *EKISTICS,* 9(55), 362–9.
18   Watts, K. 1962. How to Make an Urban Planning Survey. *EKISTICS,* 13(79), 300–12.

19   Editorial. 1960. *EKISTICS*, 9(55), 307–8.
20   Maki, F. and Goldberg, J. 1962. Linkage in Collective Form. *EKISTICS*, 14(82), 100–3.
21   Boulding, K. 1962. The Death of the City. *EKISTICS*, 13(75), 19–22.
22   The Delos Symposion. 1963. *EKISTICS*, 16(95), 205.
23   Doxiadis, C. 1963. Comment on The Delos Symposion. *EKISTICS*, 16(95), 204.
24   Psomopoulos, P. 1985. Jacky and the Delos Symposion. *EKISTICS*, 52(314/315), 493–4.
25   Editorial: Studies of Urban Form. 1962. *EKISTICS*, 15(91), 342–3.
26   The Declaration of Delos. 1963. *EKISTICS*, 16(93), n.p.
27   The CIAM Charter of Athens 1933: Outcome of a similar event. 1963. *EKISTICS*, 16 (95), 263–4.
28   First Meeting of the United Nations Committee on Housing, Building and Planning. 1963. *EKISTICS*, 15(90), 251.
29   Late career. Based in Greece, (1965–83). 1985. *EKISTICS*, 53(314/315), 407.
30   Tyrwhitt, J. 1965. The Journal of Ekistics 1955–1965: Analysis of its Contents. *EKISTICS*, 19(110), 39–44.
31   Editorial. 1966. *EKISTICS*, 22(129), 103–4.
32   Tyrwhitt, J. 1968. The Pedestrian in Megalopolis: Tokyo. *EKISTICS*, 25(147), 73–9.
33   Meier, R. 1967. Notes on the creation of an efficient megalopolis: Tokyo. *EKISTICS*, 23(138), 294–307.
34   Alexander, C. 1967. A City is not a Tree. *EKISTICS*, 23(139), 344–8.
35   Richards, J. M. 1969. Lessons from the Japanese Jungle. *EKISTICS*, 28(165), 73–4.
36   Alexander, C. 1969. Major Changes in Environmental Form Required by Social and Psychological Demands. *EKISTICS*, 28(165), 75–85.
37   Friedmann, J. 1969. Intentions and Reality: The American—Trained Planner Overseas. *EKISTICS*, 28(169), 411–15.
38   Murphy, E. F. 1985. The Skillful Editor. *EKISTICS*, 53(314/315), 455–6.
39   Editorial. 1969. *EKISTICS*, 27(160), 149–50.
40   Foreword. 1970. *EKISTICS*, 29(173), 222–4.
41   Foreword. 1971. *EKISTICS*, 31(182), 262–4.
42   Foreword. 1972. *EKISTICS*, 33(197), 234.
43   Papaioannou, J. 1985. Sparoza—or the birth of a community. *EKISTICS*, 52(314/315), 508–11.

Secondary Sources

Alofsin, A. 2002. *The Struggle for Modernism*. New York: W. W. Norton.

Anker, P. 2001. *Imperial Ecology*. Cambridge, MA: Harvard University Press.

——, 2010. *From Bauhaus to Ecohouse: A History of Ecological Design*. Baton Rouge: Louisiana State University Press.

APRR. 1941a. Broadsheet No. 4: Consumption of Fresh Food. November.

——, 1941b. Broadsheet No. 5: Production of Fresh Food. December.

——, 1942a. Broadsheet No. 6: Distribution of Fresh Food. February.

——, 1942b. The Hub of the House: Part I. The Town Kitchen. Reprint from *Architects Journal* (August 27), 1–10.

——, 1942c. Broadsheet No. 1: The Delimitation of Regions for Planning Purposes. September.

——, 1942d. Broadsheet No. 9: Regional Boundaries of England and Wales. November.

——, 1942e. Two Maps of Local Food Customs With Questionnaire. November.

——, 1942f. The Hub of the House: Part II. The Country Kitchen. December.

——, 1943. Broadsheet No. 2: General Information. April.

——, 1944a. Broadsheet No. 2: General Information. May.

——, 1944b. *Progress Sheet No. 148*. October.

——, 1945a. *Maps for the National Plan*. London: Lund Humphries.

——, 1945b. *Nature* (156)3974, 775, December 29.

——, 1946a. *An Analysis of Housing Reports 1941–45, Housing Digest*. London: Art & Educational.

——, 1946b. Plans of the Cities of Europe, from the Exhibition at the Public Museum & Art Gallery. Hastings. October.

——, 1946/47. *Information Bulletin* Sheet No. 167 December/January.

——, 1948a. *Information Bulletin* Sheet No. 182 July.

——, 1948b. *Information Bulletin* Sheet No. 183 August.

——, 1948c. *Information Bulletin* Sheet No. 184 September.

——, 1949a. *Information Bulletin* Sheet No. 193 June.

——, 1949b. *Information Bulletin* Sheet No. 198 November.

——, 1950a. *Information Bulletin* Sheet No. 202 March.

——, 1950b. *Town and Country Planning Textbook*. London: Architectural P.

Ashton, M. O. 2000. Tomorrow Town: Patrick Geddes, Walter Gropius and Le Corbusier, in *The City After Patrick Geddes*, edited by V. M. Welter and J. Lawson. Bern: Peter Lang, 191–209.

Auger, T. 1949. Comment: The Size and Spacing of Urban Communities. *Journal of the American Institute of Planners*, 15(3), 42–3.

Bailes, H. 2004. Chrystabel Prudence Procter, in *Oxford Dictionary of National Biography*, edited by H. C. G. Matthew and B. Harrison. London: Oxford University Press, 455–6.

Baker, A. 2004. James Brabazon Grimston, in *Oxford Dictionary of National Biography*, edited by H. C. G. Matthew and B. Harrison. London: Oxford University Press, 388–9.

Bauer, C. 1934. *Modern Housing*. Boston: Houghton.

——, 1948. Reconstruction. *TASK* 7/8, 3–6.

Bell, G. and J. Tyrwhitt. 1972. Introduction, in *Human Identity in the Urban Environment*, edited by G. Bell and J. Tyrwhitt. London: Penguin, 15–37.

Bevington, S. M., Blaksley, J. H., Jennings, J. R., and Tyrwhitt, J. 1939. *Leisure Pursuits Outside the Family Circle—An Account of an Investigation Carried out by the Institute's Investigators into the Leisure Pursuits Outside the Family Circle in a County Town, a Satellite Town, in a Rural District & in a Holiday Camp*. London: Natl. Inst. of Industrial Psychology.

Bickersteth, J. B. 1944. Soldiers and Citizens. *Public Opinion Quarterly*, 8(1), 10–16.

Blaksley, J. H. B. 1929. The Tory Ideal. *National Review*, 83, 539–46.

Bonham-Carter, V. 1970. *Dartington Hall The Formative Years*. Somerset: Exmoor.

Bridgwater Commission III A. 1951. Urbanism: Preparation for CIAM 7, in *A Decade of New Architecture*, edited by S. Giedion. Zurich: Girsberger, 22–5.

British Pathe. 1925. 10,000 British Fascists, Available at: http://www.britishpathe.com/video/10-000-british-fascists-aka-10-000-british-fascist/query/British+Fascisti [accessed: June 5, 2012]

Buber, M. 1937. *I and Thou*. Translated by R. G. Smith. Edinburgh: Clark.

——, 1949. *Paths in Utopia*. Translated by R. F. C. Hull. London: Routledge.

——, 1953. Introduction, in *Community and Environment: A Discourse on Social Ecology*, edited by E. A. Gutkind. New York: Philosophical, vii–ix.

Buder, S. 1990. *Visionaries and Planners: The Garden City Movement and Modern Community*. New York: Oxford University Press.

Bullock, N. 2002. *Building the Post-War World: Modern Architecture and Reconstruction in Britain*. London: Routledge.

Campbell, E. 1987. Geography at Birkbeck College, University of London, with particular reference to J. F. Unstead and E. G. R. Taylor, in *British Geography 1918–45*, edited by R. W. Steel. New York: Cambridge University Press, 45–47.

Carpenter, E. 2001. That Not so Silent Sea, in *The Virtual Marshall McLuhan*, edited by D. Theall. Montreal: McGill-Queens's University Press, 236–261.

Carpenter, E. and M. McLuhan (eds) 1960. *Explorations in Communication, an Anthology*. Boston: Beacon.

Carver, H. 1975. *Compassionate Landscape*. Toronto: University of Toronto Press.

City Planners Advise Urban Redesigning 1957. *Harvard Crimson* online edition. April 13. Available at: http://www.thecrimson.com/article.aspx?ref=108737 (accessed: June 16, 2012).

Clapson, M. 1998. *Invincible Green Suburbs, Brave New Towns*. Manchester: Manchester University Press.

Collapse of Program Feared 1954. *New York Times*, (July 25), 6.

Collins, M., F. Steiner, and M. Rushman 2001. Land Use Suitability Analysis in the United States. *Environmental Management*, 28(5), 611–21.

Conversation at CIAM 8 1952. In *The Heart of the City*, edited by J. Tyrwhitt, J. L. Sert, and E. N. Rogers. New York: Pellegrini, 36–40.

Copping, A. M. 1978. The History of the Nutrition Society. *The Proceedings of the Nutrition Society*, 37(2), 105–39.

Crang, J. A. 2000. *The British Army and the People's War 1939–1945*. Manchester: Manchester University Press.

Crinson, M. and J. Lubbock 1994. *Architecture—Art or Profession? Three Hundred Years of Architectural Education in Britain*. Manchester: Manchester University Press.

Crone, G. R. 1964. British Geography in the Twentieth Century. *Geographical Journal*, 130(2), 197–220.

Darling, E. 2007. *Re-forming Britain: Narratives of Modernity before Reconstruction*. London: Routledge.

Dartington Hall: Farming and rural crafts 1936. *The Times* (London), (April 13), 15.

Design School Offers Three New Master's Degrees in Urban Studies 1959. *Harvard Crimson*, online edition, July 16. Available at: http://www.thecrimson.com/article.aspx?ref=112579 (accessed: June 16, 2012).

Dickinson, R. E. 1938. The Economic Regions of Germany. *Geographical Review*, 28(4), 609–26.

——, 1938–39. Geography and GeoPolitik in the Third Reich. *German Life and Letters*, 3(1), 25–35.

——, 1942. The Social Basis of Physical Planning, Part I. *Sociological Review*, 34(1–2), 51–67.

Diefendorf, J. 1993. *In the Wake of War: Reconstructing German Cities after World War II*. New York: Oxford University Press.

Doebele, W. 1962. Education for Planning in Developing Countries: The Bandung School of Regional and City Planning. *Town Planning Review*, 33(2), 95–114.

Drew, J. 1945. Editor's Foreword. *Architects Year Book*, 1, 5–6.

Enfield Cable Works Ltd. 1943. Broadsheet: Vegetables and the Canteen. September.

Experimental Cottage Building 1920. *Nature*, 105(2651), 79203.

Farr, B. 1987. *The Development and Impact of Right-wing Politics in Britain 1903–1952*. New York: Garland.

Feder, G. and F. Rechenberg 1939. *Die Neue Stadt*. Berlin: Springer.

Feiss, C. 1954. Editorial. *Journal of the American Institute of Planners*, 20(4), 170–73.

Ford Foundation 1956. *Annual Report*. New York: Ford Foundation.

Gardner-Medwin, R. 1952. United Nations and Resettlement in the Far East. *Town Planning Review*, 22(4) 283–98.

Gay, J. 2006. The Path on the Hill, Sparoza. *Mediterranean Garden*, 45 (July). Available at: http://www.mediterraneangardensociety.org/journal-45.html (accessed: June 25, 2012).

Geddes, P. 1915. *Cities in Evolution*. London: Williams.

——, 1947. *Patrick Geddes in India*, edited by J. Tyrwhitt. London: Lund Humphries.

——, 1949. *Cities in Evolution*. Second edition, abridged, edited by J. Tyrwhitt. London: Williams.

——, 1968. The Valley Section, edited by J. Tyrwhitt. *Architects Year Book*, 12, 65–71.

Georgiadis, S. 1993. *Sigfried Giedion: An Intellectual Biography*. Edinburgh: Edinburgh University Press.

Gibson, T. 2011. *Brenda Colvin: A Career in Landscape*. London: Frances Lincoln.

Giedion, S, 1941. *Space, Time and Architecture: The Growth of a New Tradition*. Cambridge, MA: Harvard University Press.

——, 1948. *Mechanization Takes Command: A Contribution to Anonymous History*. New York: Oxford University Press.

——, 1950a. The Undulating Wall. *Architectural Review*, (February), 77–84.

——, 1950b. A History of the Bath. *Picture Post*, 48(1), 23–9.

——, 1951. *A Decade of New Architecture*. Zurich: Girsberger.

——, 1954a. The State of Contemporary Architecture I. The Regional Approach. *Architectural Record*, (January), 132–7.

——, 1954b. The State of Contemporary Architecture II. The Need for Imagination. *Architectural Record*, (February), 186–91.

——, 1958. *Architecture, You and Me: The Diary of a Development*. Cambridge, MA: Harvard University Press.

——, 1962. *The Eternal Present: The Beginnings of Art: A Contribution on Constancy and Change*. New York: Bollingen Foundation.

——, 1964. *The Eternal Present: Vol. 2: Beginnings of Architecture*. London: Oxford University Press.

——, 1967. *Space, Time and Architecture: The Growth of a New Tradition*. Fifth edition. Cambridge, MA: Harvard University Press.

——, 1971. *Architecture and the Phenomena of Transition: The Three Space Conceptions in Architecture*. Cambridge, MA: Harvard University Press.

Gilg, A. 1978. Policy Forum: Needed: A New 'Scott' Inquiry. *Town Planning Review*, 49(3), 353–70.

Glass, R. 1948. *The Social Background of a Plan; a Study of Middlesbrough*. London: Routledge.

Gold, J. 1997. *The Experience of Modernism*. London: Taylor.

Gottlieb, J. 2000. *Feminine Fascism*. London: I. B. Tavris.

Gould, J. 1998. Architecture in Devon, in *Going Modern and Being British*, edited by J. Smiles. Exeter: Intellect.

Gropius, W. 1935. *The New Architecture and the Bauhaus*. Translated by M. Shand. London: Faber.

Gutheim, F. 1962. The Next 50 Million Americans—Where Will They Live? *Progressive Architecture*, (August), 98–9.

Gutkind, E. A. 1943. *Creative Demobilization: Principles of National Planning.* London: Kegan Paul.

——, 1944. *Creative Demobilization: Case Studies in National Planning.* New York: Oxford University Press.

Hall, P. 1981. The Geographer and Society. *Geographical Journal,* 147(2), 145–52.

Hardy, D. 1991. *From Garden Cities to New Towns.* London: Chapman.

Hawkins, T. H. and J. F. Brimble 1947. *Adult Education–The Record of the British Army.* London: McMillan.

Heard, G. 1934. The Dartington Experiment. *Architectural Review,* (April), 119–22.

Heathorn, S. and D. Greenspoon 2006. Organizing Youth for Partisan Politics in Britain, 1918–1932. *Historian,* 68(1), 89–119.

Hebbert, M. 1983. The Daring Experiment. *Environment and Planning B: Planning & Design,* 10(1), 3–17.

Hopkinson, T. 1970. Introduction, in *Picture Post 1938–1950,* edited by T. Hopkinson. London: Chatto, 8–21.

Howard, E. 1898. *Tomorrow: A Peaceful Path to Reform.* London: Swan Sonnenschein.

Human Needs in Planning Conference at the RIBA 1946. *Journal of the Royal Institute of British Architects,* 53(February), 126–8.

Huxley, J. 1940. Book Review: Man and Society in an Age of Reconstruction. *Nature,* 146(3688), 3–4.

——, 1946a. *Unesco its Purpose and its Philosophy.* London: Preparatory Commission of The United Nations Educational Scientific and Cultural Organisation.

——, 1946b. Report of the Executive Secretary on the Work of the Preparatory Commission to the General Conference. November 20.

Ihlder, J. 1946. Replanning and Rebuilding Cities: Report on International Congress. *Planning and Civic Comment,* 12(4), 9–13.

India's Gift to Brighton 1921. *The Times* (London), (Oct 27), 7.

Jackson, P. 2010. Extremes of Faith and Nation. *Religion Compass,* 4(8), 507–17.

Jeremiah, P. 1998. A Modern Adventure, in *Going Modern and Being British,* edited by J. Smiles. Exeter: Intellect.

King, P. 1999. *Women Rule the Plot.* London: Duckworth.

Kyrstis, A. 2006. *Constantinos A. Doxiadis Texts Design Drawings Settlements.* Athens: Karos.

Koike, S. and R. Hamaguchi 1956. *Japan's New Architecture.* Tokyo: Shokokusha.

Korn, A., M. Fry, and D. Sharp 1971. The MARS Plan for London. *Perspecta,* 13, 162–73.

Kurokawa, K. 1977. *Metabolism in Architecture.* Boulder: Westview.

Lancaster, C. 1983. *The Japanese Influence in America.* New York: Abbeville.

Lanchester, H. V. 1915. The Civic Development Survey as a War Measure. *Journal of the Royal Institute of British Architects,* 22(3rd Series), 107–10.

Laurence P. 2007. Jane Jacobs Before Death and Life. *Journal of the Society of Architectural Historians*, 66(1), 5–14.

Le Lièvre, A. 1980. *Miss Willmott of Warley Place*. London: Faber.

The League of Industry: Conference at Malvern 1934. *Times* (London), (September 29), 14.

Lindsay, K. 1981. PEP through the 1930s: Organisation, Structure, People, in *Fifty years of Political & Economic Planning, Looking Forward 1931–1981*, edited by J. Pinder. London: Heinemann, 9–31.

Linehan, T. 2000. *British Fascism, 1918–1939: Parties, Ideologies and Culture*. Manchester: Manchester University Press.

Mallgrave, H. F. 2005. *Modern Architectural Theory: A Historical Survey, 1673–1968*. Cambridge, UK: Cambridge University Press.

Mandle, W. F. 1967. Sir Oswald Leaves the Labour Party, March 1931. *Labour History*, 12, 35–51.

Mannheim, K. 1940. *Man and Society in an Age of Reconstruction*. London: Routledge.

——,1943. Sociology for the Educator and the Sociology of Education, in *Sociology and Education*, Le Play House, Malvern [addresses given at Winter School of Sociology and Civics organized by the Institute of Sociology [Le Play House] at St. Hilda's College, Oxford, January, 4–9.

Mantziaras, P. 2003. Rudolf Schwartz and the Concept of Standtlandschaft. *Planning Perspectives*, 18(2), 147–176.

Marshall, R. 2008. Shaping the City of Tomorrow, in *Josep Lluis Sert: The Architect of Urban Design, 1953–1969*, edited by E. Mumford and H. Sarkis. New Haven: Yale University Press, 130–43.

McAtee, C. 2008. From the Ground Up, in *Josep Lluis Sert: The Architect of Urban Design, 1953–1969*, edited by E. Mumford and H. Sarkis. New Haven: Yale University Press, 166–97.

McCallum, I. R. M. 1945. *Physical Planning, The Groundwork of a New Technique*. London: Architectural P.

McGuken, W.1984. *Scientists, Society and the State*. Columbus: Ohio State University Press.

Mead, M. 1978. Interview, in *Dialogues with Delians*, edited by M. Perovic. Ljubljana: Sintenza, TK.

Meech, J. 1988. Japonisme at the Turn of the Century, in *Perspectives on Japonisme: The Japanese Influence in America*, edited by P. D. Cate. Zimmerli Art Museum: Rutgers University Press, 19–33.

Meller, H. 1990. *Patrick Geddes: Social Evolutionist and City Planner*. New York: Routledge.

Meyerson, M. with J. Tyrwhitt, B. Falk, and O, Seklerl. 1963. *Face of the Metropolis*. New York: Random.

Ministry of Town and Country Planning 1950. Report of the Committee on Qualifications of Planners, (Cmd. 8059). London: HMSO.

Misc. Notes 1934. *Bulletin of Miscellaneous Information*. Royal Gardens, Kew, 39(9), 398.

Low Cost Housing in South and South East Asia 1951. Report of Mission of Experts. July. New York: United Nations.

Moholy-Nagy, L. 1945. Introduction, in W. Gropius, *Rebuilding Our Communities*. Chicago: Paul Theobold, 11–12.

Mowat, C. L. 1967. *Britain Between the Wars, 1918–1940*. Chicago: University of Chicago Press.

Mumford, E. 2000. *The CIAM Discourse on Urbanism, 1928–1960*. Cambridge, MA: MIT Press.

——, 2009. *Defining Urban Design, CIAM Architects and the Formation of a Discipline, 1937–69*. New Haven: Yale University Press.

Mumford, E. and H. Sarkis 2008. *Josep Lluis Sert: The Architect of Urban Design, 1953–1969*. New Haven: Yale UP.

Mumford, L. 1934. *Technics & Civilization*. New York: Harcourt.

——, 1938. *The Culture of Cities*. New York: Harcourt.

——, 1946. Address given at a General Meeting of the Town Planning Institute, June 20, 1946. *Journal of the Town Planning Institute*, July-August, 1–4.

——, 1947. Introduction, in *Patrick Geddes in India*, edited by J. Tyrwhitt. London: Lund Humphries, 7–13.

——, 1950. Mumford on Geddes. *Architectural Review*, 108(644), 81–7.

Nicholson, M. 1981. PEP through the 1930s: Growth, Thinking, Performance, in *Fifty Years of Political & Economic Planning, Looking Forward 1931–1981*, edited by J. Pinder. London: Heinemann, 32–53.

Osborn, F. 1946. To L. Mumford, in *The Letters of Lewis Mumford and Frederick Osborn: A Transatlantic Dialogue, 1938–1970*, edited by L. Mumford and F. Osborn. New York: Praeger, October 17.

Otto, C. 1965. City Planning Theory in Nationalist-Socialist Germany. *Journal of the Society of Architectural Historians*, 24(1), 70–74.

Parmar, I. 2002. American Foundations and the Development of International Knowledge Networks. *Global Networks*, 2(1), 13–30.

Pearse, I. 1943. *The Peckham Experiment: A Study in the Living Structure of Society*. London: Allen.

Pepler, G. 1923. International Notation for Civic Surveys and Town Plans. International Garden Cities and Town Planning Federation Conference Papers, Report No. 5. Gothenburg.

Planning of Metropolitan Areas and New Towns 1969. Meeting of the United Nations Group of Experts on Metropolitan Planning and Development, Stockholm 14–30 September 1961; United Nations Symposium on the Planning and Development of New Towns, Moscow, 24 August–7 September 1964. New York: United Nations.

Planners to Investigate 'Aesthetic Urban Design' 1956. *Harvard Crimson*, online edition January 13. Available at: http://www.thecrimson.com/article.aspx?ref=214175 (accessed: June 16, 2012).

Powers, A. 2002. Landscape in Britain, in *The Architecture of Landscape 1940–1960s*, edited by M. Treib. Philadelphia: University of Pennsylvania Press, 56–81.

Presthus, R.V. 1951. The Schuster Report: An Interpretation. *Journal of the American Institute of Planners*, 17(1), 43–5.

Procter, T. 2009. *Scouting for Girls*. Santa Barbara: Praeger.

Protection for Agriculture 1934. *Times* (London), (September 25), 11.

Pugh, M. 2005. *Hurrah for the Blackshirts*. London: Jonathan Cape.

Pyla, P. 2008. Back to the Future: Doxiadis' Plans for Baghdad. *Journal of Planning History*, 7(1), 3–19.

Rayward, W. B. 1994. Visions of Xanadu: Paul Otlet (1868–1944) and Hypertext. *Journal of the American Society for Information Science*, 45(4), 235–250.

Read, H. 1943. Foreword, in E. A. Gutkind, *Creative Demobilization Volume I: Principles of National Planning*. London: Kegan Paul, xiii–xvi.

Resolutions 1952. *Economic and Social Council Official Records: Fourteenth Session 20 May–1 August 1952 Supplement* 1. New York: United Nations.

Rice, D. 1969. UK's Counter-insurgency Work in Indonesia and Thailand: An Expose. *Blue Tail Fly*, (8), 5–7, 17–18.

Ritschel, D. 1997. *The Politics of Planning*. Oxford: Clarendon.

Roosmalen, P. K. M. van 2005. Expanding Grounds. The Roots of Spatial Planning in Indonesia, in *The History of the Indonesian City before and after Independence*, edited by F. Colombijn and M. Barwegen. Yokyakarta: Ombak, 75–117.

Roskill, O. 1981. PEP through the 1930s: The Industries Group, in *Fifty years of Political & Economic Planning, Looking Forward 1931–1981*, edited by J. Pinder. London: Heinemann, 54–80.

Ross, M. F. 1978. *Beyond Metabolism: The New Japanese Architecture*. New York: Architectural Record.

Ross, E. 2007. *Slum Travelers*. Berkeley: University of California Press.

Rowse, E. A. A. 1939. The Planning of a City. *Journal of the Town Planning Institute*, 25(March), 167–71.

Sackville-West, V. 1944. *The Women's Land Army*. London: Michael Josephy.

Saunier P.-Y. 2001. Sketches from the Urban Internationale, 1910–50: Voluntary Association, International Institutions and US Philanthropic Foundations. *International Journal of Urban and Regional Research*, 25(2), 380–403.

Schenk, T. and R. Bromley 2003. Mass Producing Traditional Small Cities: Gottfried Feder's Vision for a Greater Nazi Germany. *Journal of Planning History*, 2(2), 107–39.

Science and world order. 1941. *The Times* (London), (September 26), 5.

Sert, J. L. 1942. *Can Our Cities Survive?* Cambridge, MA: Harvard University Press.

——, 1956. Graduate School of Design, in Report to the President of Harvard College and Reports of Departments, 492–501. Harvard University, Harvard

University Archives. Available at: http://pds.lib.harvard.edu/pds/view/2582287 (accessed: June 16, 2012).

——, 1957. Graduate School of Design, in Report to the President of Harvard College and Reports of Departments, 425–34. Harvard University, Harvard University Archives. Available at: http://pds.lib.harvard.edu/pds/view/2582287 (accessed: June 16, 2012).

Sheail, J. 1997. Scott Revisited: Post-war Agriculture, Planning and the British Countryside. *Journal of Rural Studies*, 13(4), 387–98.

Short Outlines of the Core 1952. In *The Heart of the city: Towards the Humanisation of Urban Life*, edited by J. Tyrwhitt, J. L. Sert, and E. N. Rogers. New York: Pellegrini, 164–8.

Shoshkes, E. 2004. East-West: Interactions between the United States and Japan and their Effect on Utopian Realism. *Journal of Planning History*, 3(3), 215–40.

Smiles, J. 1998. Refuge or Regeneration, in *Going Modern and Being British*, edited by J. Smiles. Exeter: Intellect.

——, 1998. *Going Modern and Being British*. Exeter: Intellect.

Stead, H. G. 1942. *The Education of a Community, Today and Tomorrow*. London: University of London.

Sullivan, M. 1989. *The Meeting of Eastern and Western Art*. Berkeley: University of California Press.

Tansley, A. 1922. *Elements of Plant Biology*. London: Allen.

——, 1923. *Practical Plant Biology*. London: Allen.

Taper, B. 1967. A Lover of Cities Part I. *New Yorker* (February 4), 39–91.

Taylor, E. G. R. 1945. *Planning Prospect Survey before Plan series no. 1*. London: Lund Humphries.

Taylor, W. 1949. Review, Planning 1948. *Land Economics*, 25(3), 328–9.

Thomas, A. 1996. *Portraits of Women*. Cambridge, UK: Polity.

Thomas, D. 1963. London's Green Belt. *Geographical Journal*, 129(1), 14–24.

Thomas, C. P. 1959. A Great Benefaction. *Proceedings of the Royal Society of Medicine*, 53 (January), 1–8.

Thomas, M. H. 1951. Report of CIAM 6 Bridgwater 1947, in *A Decade of New Architecture*, edited by S. Giedion. Zurich: Girsberger, 8–11.

Thurlow, R. 1987. *Fascism in Britain, a History: 1918–1935*. Oxford, UK: Blackwell.

Turner, J. 2000. Interview Transcript. World Bank September 11. Available at: http://siteresources.worldbank.org/INTUSU/Resources/turner-tacit.pdf (accessed: June 12, 2012).

Twinch, C. 1990. *Women of the Land: Their Story during Two World Wars*. Cambridge, UK: Lutterworth.

Tyrwhitt, J. 1945. Town Planning. *Architects Year Book*, 1, 11–29.

——, 1946a. Training the Planner, in *Planning and Reconstruction*, edited by T. Todd. London: Todd, 209–13.

——, 1946b. *Planning and the Countryside*. London: Art & Educational.

——, 1947a. *Patrick Geddes in India*. London: Lund Humphries.

——, 1947b. Training the Planner in Britain. *International Federation of Housing and Town Planning News Sheet* (December), 209–13.

——, 1948a. Reconstruction: Great Britain. *TASK* 7/8, 20–24.

——, 1948b. Opening up Greater London. *National Municipal Review*, 37(11), 592–6.

——, 1949a. Introduction, in P. Geddes, *Cities in Evolution*, Second edition, abridged, edited by J. Tyrwhitt. London: Williams, ix–xxvii.

——, 1949b. The Size and Spacing of Urban Communities. *Journal of the American Institute of Planners*, 15(2), 10–17.

——, 1949c. The CIAM Grid. *APRR Information Bulletin* Sheet No. 198. November, 1.

——, 1950a. Preface, in *Town and Country Planning Textbook*, edited by APRR. London: Architectural Press, xv–xvi.

——, 1950b. Society and Environment. A Historical Review, in *Town and Country Planning Textbook*, edited by APRR. London: Architectural Press, 96–145.

——, 1950c. Surveys for Planning, in *Town and Country Planning Textbook*, edited by APRR. London: Architectural Press, 146–278.

——, 1951a. The Valley Section, Patrick Geddes' World Image. *Journal of the Town Planning Institute*, 37(3), 61–6.

——, 1951b. Do New Towns Provide Safety? No. *Progressive Architecture*, 32(9), 77.

——, 1954. Ideal Cities and the City Ideal. *Explorations 2*, 38–50.

——, 1955a. The Moving Eye. *Explorations 4*, 90–5.

——, 1955b. Chandigarh. *Journal of the Royal Architectural Institute of Canada*, 32(353), 11–20.

——, 1955c. The City Unseen. *Explorations 5*, 88–96.

——, 1958. Fatehpur Sikri. *Architectural Review*, 123(733), 124–8.

——, 1959. The Civic Square. *Canadian Architect*, (April), 55–65.

——, 1962. Education for Urban Design. *Journal of Architectural Education*, 17(3), 100–102.

——, 1963. Shapes of Cities That Can Grow. *Architectural Association Journal*, 79(876), 87–102.

——, 1965. The Ekistic Grid. *Architectural Association Journal*, 81(894), 10–15.

——, 1966a. Education for Urban Design: Origin and Concepts of the Harvard Program, in *The Architect and the City*, edited by M. Whiffen. Cambridge, MA: MIT Press.

——, 1966b. Chios. *Architectural Review*, 139(832), 475–8.

——, 1968. Chios. *Architects Year Book*, 12, 194–230.

——, 1978. Interview, in *Dialogues with Delians*, edited by M. Perovic. Ljubljana: Sintenza, 127–48.

——, 1980. Working with Charles Abrams. *Habitat International*, 5(1/2), 35–6.

——, 1998. *Making a Garden on a Greek Hillside*. Limni: Denise Harvey.

Tyrwhitt, R. P. 1862. Notices and Remains of the Family of Tyrwhitt. N.p.

Tyrwhitt, J. and B. Colvin. 1947. *Trees for Town and County*. London: Lund Humphries.

Tyrwhitt, J. and W. L. Waide. 1949. *Basic Surveys for Planning*. London: Architect and Building News.

Tyrwhitt J. J. L. Sert and E. N. Rogers 1952. *The Heart of the City: Towards the Humanisation of Urban Life*. New York: Pellegrini.

United Nations Educational, Scientific and Cultural Organization 1947. *International Organization*, 1(1), 130–33.

Urban Design 1956. *Progressive Architecture*, 37(8), 97–112.

Van den Heuvel, C. 2008. Building Society, Constructing Knowledge, Weaving the Web: Otlet's Visualizations of a Global Information Society and His Concept of a Universal Civilization, in *European Modernism and the Information Society*, edited by B. Rayward. London: Ashgate, 127–53.

Ward, S. 2004. *Planning and Urban Change*. London: Sage.

——, 2010. What Did the Germans Ever Do For Us? A Century of British Learning About and Imagining Modern Town Planning. *Planning Perspectives*, 25(2), 117–40.

Watts, K. 1997. *Outward from Home: A Planner's Odyssey*. Lewes: Book Guild Ltd.

Webster, D. 2011. Development Advisors in a Time of Cold War and Decolonization: The United Nations Technical Assistance Administration, 1950–59. *Journal of Global History*, 6, 249–72.

Weissmann, E. 1978. Human Settlements—Struggle for Identity. *Habitat International*, 3(314), 227–41.

Wehner, B. 1937. German Regional Planning. *Journal of the American Institute of Planners*, 3(1), 9–12.

Welter, V. 2002. Post-War CIAM, Team X and the Influence of Patrick Geddes. Available at: http://www.Team10online.org/research/studies_and_papers.html (accessed: June 25, 2012).

Wersky, G. 1978. *The Visible College*. New York: Holt.

White, B. 1974. *The Literature and Study of Urban and Regional Planning*. London: Routledge.

Willatts, E. C. 1987. Geographers and Their Involvement in Planning, in *British Geography 1918–45*, edited by R. W. Steel. New York: Cambridge University Press, 100–15.

Wilson, E. 2003. *Bohemians*. New Brunswick: Rutgers University Press.

Young, M. 1982. *The Elmhirsts of Dartington*. London: Routledge.

Index